Cody's Collection of Popular
SAS® Programming Tasks
and How to Tackle Them

Ron Cody

support.sas.com/bookstore

The correct bibliographic citation for this manual is as follows: Cody, Ron. 2012. *Cody's Collection of Popular SAS® Programming Tasks and How to Tackle Them*. Cary, NC: SAS Institute Inc.

Cody's Collection of Popular SAS® Programming Tasks and How to Tackle Them

SAS Institute Inc. provides a complete selection of books and electronic products to help customers use SAS software to its fullest potential. For more information about our e-books, e-learning products, CDs, and hard-copy books, visit the SAS Books Web site at **support.sas.com/bookstore** or call 1-800-727-3228.

Contents

List of Programs

Chapter 4

Chapter 8

Chapter 9

Chapter 10

Chapter 11

xiv

About This Book

Purpose

There are quite a few programming tasks that, at some point in your programming career, you need to accomplish. For example, you may need to convert character values to numeric values or combine summary data with detail data. Why "reinvent the wheel" every time you need to accomplish one of these tasks? That's what this book is all about—it contains annotated programs to solve many of the common programming tasks you face on a daily basis.

Is This Book for You?

This book is useful for all levels of programmers, from novice to expert.

Prerequisites

To benefit from this book, you need to be able to write SAS programs, using either the Base SAS Enhanced Editor or SAS Enterprise Guide.

Scope of This Book

There are 12 chapters in this book, covering many of the more popular programming tasks. In many cases, the book presents a program to solve a task as well as a macro that performs the task automatically. You will find several macros in the book that will save you hours of tedious programming. Many of the tasks, such as determining the number of observations in a SAS data set, are solved using four or five different approaches. This allows you to choose the method you prefer as well as providing a "teaching moment" to explore programming techniques you may not have seen before.

Highlights

Were you ever given a SAS data set with variables such as Height, Weight, or Age, and they were all character values? If so, join the club. One of the tasks I include in this book shows you an easy way to convert selected character variables into numeric variables. Better still, I include a macro that allows you to specify an input and output data set name, along with a list of character variables that need converting, and the macro automatically creates a new data set with all the character variables converted to numeric variables, maintaining the original variable names.

Do you ever need to check for possible errors in your numeric values? If so, you will find two macros in this book that will help you. One of them allows you to specify ranges for each numeric variable and generate an error report for any observations where a value is outside the specified range. The other macro uses automatic outlier detection to report possible data errors.

Have you ever needed to concatenate two SAS data sets? Of course you have! This is easily accomplished using a SET statement or PROC APPEND. What if the two data sets have character variables whose lengths are not the same in the two data sets? This problem is solved by a macro that concatenates the two data sets, using the longer length of any character variables that are not the same in the two data sets.

It is sometimes convenient to perform an operation on every numeric variable or every character variable in a SAS data set. For example, you may want to convert all numeric values of 999 to a SAS missing value, or you may want to convert all character variables to uppercase. These tasks are demonstrated in this book.

How about combining summary data (perhaps a mean of all observations) with detail data. One of the chapters shows you several interesting ways to accomplish such tasks.

Restructuring (also called transposing) a data set from one observation to many or vice versa is a common programming requirement. You will find programs to perform these tasks using a DATA step approach and using PROC TRANSPOSE.

Author Page

You can access this book's author page at support.sas.com/cody.

This author page includes all of the SAS Press books that this author has written. For features that relate to this specific book, look for the cover thumbnail of this book. The links below the book will take you to a free chapter, example code and data, reviews, updates, and more.

Example Code and Data

You can access the example code and data for this book by linking to its author page at support.sas.com/cody. Look for the cover thumbnail of this book, and select Example Code and Data to display the SAS programs that are included in this book.

For an alphabetical listing of all books for which example code and data is available, see support.sas.com/bookcode. Select a title to display the book's example code.

If you are unable to access the code through the Web site, send e-mail to saspress@sas.com.

To Contact the Author through SAS Press

By mail:

SAS Institute Inc.
SAS Press
Attn: Ron Cody
SAS Campus Drive
Cary, NC 27513-2414

By e-mail: saspress@sas.com

Via the Web: support.sas.com/author_feedback

SAS Books

For a complete list of books available through SAS, visit support.sas.com/bookstore.

Phone: 1-800-727-3228

Fax: 1-919-677-8166

E-mail: sasbook@sas.com

SAS Book Report

Receive up-to-date information about all new SAS publications via e-mail by subscribing to the SAS Book Report monthly eNewsletter. Visit support.sas.com/sbr.

About The Author

Ron Cody, EdD, is a retired professor from the Robert Wood Johnson Medical School who now works as a private consultant and a national instructor for SAS. A SAS user since 1977, Ron's extensive knowledge and innovative style have made him a popular presenter at local, regional, and national SAS conferences. He has authored or co-authored numerous books.

Learn more about this author by visiting his author page at support.sas.com/cody. There you can download free chapters, access example code and data, read the latest reviews, get updates, and more.

Acknowledgments

It seems that with each book I write, more and more people are needed to move the book from concept to final product. In the "old days," SAS Press only needed to produce a book—on paper. Today, books are made available on a wide variety of devices such as Kindles, iPads, etc., making for a lot of extra work.

After all the chapters are complete, a team of reviewers, some from SAS Institute, some from outside of SAS, read all or part of the book for technical errors. This group of technical reviewers did a splendid job of spotting programming errors (who, me?) and sometimes suggesting some clever and elegant solutions for some of my tasks. My sincere thanks to the following technical reviewers: Amber Elam, Paul Grant, Mark Jordan, Bari Lawhorn, Jerry Leonard, Kathryn McLawhorn, Ginny Piechota, Jan Squillace, Russ Tyndall, and Kim Wilson.

This is the third book that I have written with John West as the pivotal person, coordinating the many tasks needed to produce a book. John, thanks so much. Special thanks also go to Mary Beth Steinbach, who volunteered to perform the copy edit. Other SAS staff who worked on my book, whom I would like to thank are:

Jennifer Dilley, graphics design and back cover design
Candy Farrell, technical publishing specialist
Luis Franco, front cover design
Stacey Hamilton, marketing
Julie Platt, editor in chief
Aimee Rodriguez, marketing

It is such a pleasure writing books with the support of such talented and dedicated people. Thank you SAS Press!

Chapter 1: Tasks Involving Conversion: Character to Numeric, Specific Values to Missing, and Changing Case

Introduction

This chapter contains programs to perform character-to-numeric conversion, one of the most common tasks you will face as a SAS programmer. You will see a sample program as well as a useful macro that accomplishes this goal.

Another task that you will probably face is converting a specific numeric value such as 999 or a specific text value such as 'NA' to a SAS missing value.

In this chapter, you will also see how to convert every character variable to a specific case, such as uppercase.

The last task in this chapter demonstrates how to read data values that contain units, such as 100Lbs. or 50Kgs. and create a numeric variable with all of the values using the same units.

Task: Converting character values to numeric values

Keywords

Character-to-numeric conversion

Swap and Drop

How many times have you been given a SAS data set with variables such as Height or Weight but, instead of being numeric variables, they are stored as character? The following example describes how to convert these character variables to numeric variables, maintaining the original variable names.

For this example, you start out with a SAS data set called Char_values. Here is a listing:

Age	Weight	Gender	DOB
23	150	M	10/21/1983
67	220	M	09/12/2001
77	101	F	05/06/1977

If you run PROC CONTENTS on this data set, you see that Age and Weight are character variables. The following program performs the conversion:

Program 1.1: Converting character values to numeric values

```
*Converting character values to numeric;

data Num_Values;
   set Char_Values(rename=(Age = C_Age
                           Weight = C_Weight));
   Age = input(C_Age,best12.);
   Weight = input(C_Weight,best12.);
   drop C_:;
run;
```

The "trick" here is to rename the variables as they are read from the input data set. This allows you to use the original variable names for the resulting numeric variables. The character-to-numeric conversion is performed using the INPUT function. You don't have to worry if the INFORMAT used in the INPUT function represents more digits than you need—unlike an INPUT statement, you can never read past the end of a character value when using the INPUT function.

Notice the variable list on the DROP statement C_: The colon acts as a wildcard suffix. C_: represents all variables that begin with the characters C followed by an underscore.

The resulting data set has exactly the same variables as the original data set except the two variables Age and Weight are now numeric. A partial listing from PROC CONTENTS confirms this:

Alphabetic List of Variables and Attributes			
#	Variable	Type	Len	Format
3	Age	Num	8	
2	DOB	Num	8	MMDDYY10.
1	Gender	Char	1	
4	Weight	Num	8	

Task: Converting character values to numeric values using a macro

Keywords

Character-to-numeric conversion

Conversion macro

Because character-to-numeric conversion is required in so many situations, this chapter offers you a macro that performs the conversion automatically. As in the previous program, the resulting data set uses the same variable names as in the original data set that contains the character variables. Here is the macro, followed by an explanation:

Program 1.2: Presenting a macro to perform character-to-numeric conversion

```
*Macro to convert selected character variables to
 numeric variables;
%macro char_to_num(In_dsn=,    /*Name of the input data set*/
               Out_dsn=,   /*Name of the output data set*/
               Var_list=  /*List of character variables that you
                             want to convert from character to
                             numeric, separated by spaces*/);
   /*Check for null var list */
   %if &var_list ne %then %do;
   /*Count the number of variables in the list */
   %let n=%sysfunc(countw(&var_list));
   data &Out_dsn;
      set &In_dsn(rename=(
      %do i = 1 %to &n;
```

```
      /* break up list into variable names */
         %let Var = %scan(&Var_list,&i);
      /*Rename each variable name to C_ variable name */
         &Var = C_&Var
      %end;
      ));

   %do i = 1 %to &n;
      %let Var = %scan(&Var_list,&i);
      &Var = input(C_&Var,best12.);
   %end;
   drop C_:;
   run;
   %end;
%mend char_to_num;
```

The calling arguments in this macro are the names of the input and output data sets and a list of the variables that you wish to convert from character to numeric. You enter the names of each variable in this list, separated by spaces.

The first task of the macro is to rename each of the original variable names by appending the prefix C_ to each of the names. To determine how many variable names there are in &Var_list, you use the COUNTW function. This function computes the number of words in a string. To obtain each of the variable names, you use the %SCAN macro function. This functions works in the same way as the regular non-macro SCAN function. The first argument is the list of variable names. The second argument specifies which "word" you want in the string. The macro uses a %DO loop to extract each of the individual variable names.

The next %DO loop performs the character-to-numeric conversion using the INPUT function. Notice that the first argument of the INPUT function is the original variable name with the C_ prefix added. Finally, a DROP statement deletes all of the C_ variables.

To test the macro, you can use the original data set Char_values and enter Age and Weight as the argument of Var_List. Here is the code:

Program 1.3: Testing the character-to-numeric conversion macro

```
*Test the macro;
%char_to_num(In_dsn=char_values, Out_dsn=Num_values,
             Var_list=Age Weight)
```

After you run the macro, the output data set (Num_values) is identical to the one created by the previous program.

Task: Converting a specific value such as 999 to a missing value for all numeric variables in a SAS data set

Keywords

Numeric variables

numeric

Array

You will find numerous occasions where you need to perform an operation on all numeric (or character) variables in a SAS data set. For example, you may have a SAS data set where specific values, such as 999 or 9999, were used to represent a missing value. In the character domain, you may want to convert all character values to uppercase or convert a specific value such as 'NA' to a SAS missing value. The approach to all of these tasks is the same. You create an array of all numeric or character variables. Once you do this, you can then use a DO loop to perform any operation you desire on all of the variables in the array.

This first example converts a value of 999 to a SAS missing value for all the numeric variables in data set Demographic.

A listing of data set Demographic is shown here:

Subj	Score	Weight	Heart_Rate	DOB	Gender	Party
1	70	999	76	04NOV1955	Male	NA
2	26	160	62	08APR1955	NA	NA
3	71	195	71	20JUL1955	male	na
4	40	132	74	08JAN1955	Male	Republican
5	999	181	62	15AUG1951	Female	Democrat
6	62	71	52	24JAN1950	Male	democrat
7	24	136	72	26NOV1950	Female	democrat
8	5	174	71	08NOV1950	Female	democrat
9	5	172	47	28DEC1951	Male	Democrat
10	94	173	999	06MAY1953	Male	republican
11	99	170	63	27FEB1950	na	NA
12	10	133	63	18MAR1954	Male	democrat
13	6	131	60	26MAR1951	Female	republican

Subj	Score	Weight	Heart_Rate	DOB	Gender	Party
14	999	140	79	01OCT1950	NA	na
15	999	124	999	12OCT1950	NA	na
16	44	194	72	31DEC1952	Female	republican
17	62	196	68	09MAR1951	Female	democrat
18	57	133	72	15SEP1951	Female	Democrat
19	45	137	86	16NOV1951	NA	Republican
20	90	170	80	01OCT1951	Female	Republican

You will use this data set for several of the tasks in this chapter. For this example, notice that there are several values of 999 for the variables Score, Weight, and Heart_Rate.

Here is the code that performs the conversion:

Program 1.4: Converting a specific value such as 999 to a missing value for all numeric variables in a SAS data set

```
*Converting a specific value such as 999 to a missing value for
 all numeric variables in a SAS data set;

data Num_missing;
   set Demographic;
   array Nums[*] _numeric_;
   do i = 1 to dim(Nums);
      if Nums[i] = 999 then Nums[i] = .;
   end;
   drop i;
run;
```

The key to this program, as well as several programs to follow, is to create an array using the keyword _NUMERIC_. When used in a DATA step, _NUMERIC_ represents all the numeric variables that have been defined up to that point in the DATA step. Since the ARRAY statement follows the SET statement, the Nums array contains all of the numeric variables in data set Demographic (Subj, Score, Heart_Rate, and DOB). To make this important point clear, had you placed the ARRAY statement before the SET statement, the array Nums would not contain any variables.

You certainly do not want to have to count all the numeric variables in a large data set. Therefore, you use an asterisk in the brackets following the array name. When you do this, SAS will count the number of variables for you. But, what value do you use in the DO loop? You can use the DIM (dimension) function to determine how many variables are in the array. Your work is almost finished. All you need to do now is to check for values of 999 and convert them to a SAS numeric missing value. Don't forget to drop the DO loop counter.

The first five observations in data set Num_missing are shown next, to demonstrate that the program worked as expected:

Subj	Score	Weight	Heart_Rate	DOB	Gender	Party
1	70	.	76	04NOV1955	Male	NA
2	26	160	62	08APR1955	NA	NA
3	71	195	71	20JUL1955	male	na
4	40	132	74	08JAN1955	Male	Republican
5	.	181	62	15AUG1951	Female	Democrat

Task: Converting a specific value such as 'NA' to a missing value for all character variables in a SAS data set

Keywords

Character variables

character_Array

This task is similar to the previous task. The difference is that you want to convert a specified character value to a SAS character missing value. All you need to do is use the SAS keyword _CHARACTER_ to create an array of all character variables. Here is the program:

Program 1.5: Converting a specific value such as 'NA' to a missing value for all character variables in a SAS data set

```
*Converting a specific value such as "NA" to a missing value for all
 character variables in a SAS data set;
data Char_missing;
   set Demographic;
   array Chars[*] _character_;
   do i = 1 to dim(Chars);
      if Chars[i] in ('NA' 'na') then Chars[i] = ' ';
   end;
   drop i;
run;
```

Array Chars contains all the character variables in data set Demographic (in this case, Gender and Party). As in the previous task, the DIM function returns the number of variables in the array. To make the program more general, it looks for uppercase or lowercase values of 'NA'. Here is a listing of the first five observations in data set Char_missing:

Subj	Score	Weight	Heart_Rate	DOB	Gender	Party
1	70	999	76	04NOV1955	Male	
2	26	160	62	08APR1955		
3	71	195	71	20JUL1955	male	
4	40	132	74	08JAN1955	Male	Republican
5	999	181	62	15AUG1951	Female	Democrat

Task: Changing all character values to either uppercase, lowercase, or proper case

Keywords

Uppercase

Lowercase

Proper case

character

In a similar manner to the previous program, you can use an array of all your character variables to convert them all to a unified case: uppercase, lowercase, or proper case. Please refer to the previous program if you would like an explanation of this program. As you can see, this program is converting all the character values in the Demographic data set to uppercase. The two other functions that convert character values to lowercase or proper case are LOWCASE and PROPCASE, respectively. Here is the program:

Program 1.6: Changing case for all character variables in a SAS data set

```
*Converting all character values to uppercase (or lower- or proper-
case);
 data Upper;
   set Demographic;
   array Chars[*] _character_;
   do i = 1 to dim(Chars);
      Chars[i] = upcase(Chars[i]);
   end;
   drop i;
run;
```

If the character variables you are dealing with represent names and addresses, after you have converted all the values to a consistent case, you may want to take the additional step and use the COMPBL function to convert all multiple blanks to a single blank, to help standardize the names and addresses.

Task: Reading a numeric value that contains units such as Lbs. or Kgs. in the value

Keywords

Character-to-numeric conversion

Removing units from a value

Extracting digits from a string

COMPRESS function

SCAN function

Data set Units contains a character variable called Weight that includes units such as Lbs. and Kgs. (pounds and kilograms). To add insult to injury, the variable Height also contains units and it is expressed in feet and inches (sometimes the inches value is missing (when the inches value is zero). A listing of data set Units is shown here:

Subj	Weight	Height
001	80kgs	5ft 3in
002	190lbs	6' 1"
003	70KG.	5ft 11in
004	177LbS.	5' 11"
005	100kgs	6ft

Notice that the Weight units are not always in the same case and some of the units end in periods. For Height, the abbreviation 'ft' or 'in' is used; sometimes a single quote and double quote represent feet and inches.

You would like to create two new variables (Weight_Lbs and Height_Inches) that are numeric variables and are equal to the weight in pounds and the height in inches, respectively. Here is the program:

Program 1.7: Reading data values that contain units

```
*Reading data values that contain units;
 data No_Units;
    set Units;
    Weight_Lbs = input(compress(Weight,,'kd'),12.);
    if findc(Weight,'k','i') then Weight_lbs = Weight_lbs*2.2;
    Height = compress(Height,,'kds');
    Feet = input(scan(Height,1,' '),12.);
    Inches = input(scan(Height,2,' '),12.);
```

```
      if missing(Inches) then Inches = 0;
      Height_Inches = 12*Feet + Inches;
      drop Feet Inches;
run;
```

You start by extracting the digits from Weight using the COMPRESS function with the modifiers 'kd' (keep digits). It is important to include two commas following the first argument of the COMPRESS function so that the function interprets 'kd' as modifiers and not the second argument to the COMPRESS function that is used to list the characters you want to compress from a string. Since the result of the COMPRESS function is a character value, you use the INPUT function to perform the character-to-numeric conversion. All you need to do is test the original variable (Weight) to see if it contains a 'K' in uppercase or lowercase. Use the FINDC function with the 'i' modifier (ignore case) to do this. If you find a 'K', you multiply by 2.2 to convert from kilograms to pounds.

The Height variable presents more of a challenge. You first use the COMPRESS function with three modifiers, 'kds' (keep digits and space characters). The variable Height now contains two sets of digits (or only a single digit if there are zero inches) and can use the SCAN function to extract the feet and inch values. The SCAN function returns a missing value for Inches if Height only contains a single number (feet). You can now add 12 times the feet plus the number of inches to obtain the height in inches. Here is the listing of the data set No_Units:

Subj	Weight	Height	Weight_Lbs	Height_Inches
001	80kgs	5 3	176	63
002	190lbs	6 1	190	73
003	70KG.	5 11	154	71
004	177LbS.	5 11	177	71
005	100kgs	6	220	72

Solving this task without the COMPRESS and SCAN functions would certainly be a challenge—with these functions, it's a snap.

Task: Solving part of the previous task using a Perl regular expression

Keywords

Removing units from a value

Extracting digits from a string

Perl regular expression

My younger son, who is a wizard at programming, suggested I solve this problem using a Perl regular expression. This solution is not simpler than the previous solution, but it demonstrates the versatility of regular expressions.

You start by using PRXPARSE to compile the regular expression:

```
/^(\d+)(\D)/
```

This regex (this is what Perl programmers call regular expressions) is looking for one or more digits followed by a non-digit. The ^ in the beginning of the expression says to start the search at the beginning of the string. The digit and non-digit values will be placed in capture buffers because each of these expressions is in a set of parentheses. You use the PRXMATCH function to search for the pattern of a number followed by a non-number. The PRXPOSN function extracts the values in each of the capture buffers. The INPUT function performs the character-to-numeric conversion as in the previous task.

If the value in the second capture buffer is a 'K', you perform the kilogram to pound conversion.

Program 1.8: Using a Perl regular expression to extract the digit and units part of a character value

```
*Solution using Perl Regular expressions;
data No_Units;
   set Units(drop=Height);
   if _n_ = 1 then do;
      Regex = "/^(\d+)(\D)/";
      re = prxparse(Regex);
   end;
   retain re;
   if prxmatch(re,Weight) then do;
      Weight_Lbs = input(prxposn(re,1,Weight),8.);
      Units = prxposn(re,2,Weight);
      if upcase(Units) = 'K' then Weight_Lbs = Weight_Lbs*2.2;
   end;
   keep Subj Weight Weight_Lbs;
run;
```

The resulting data set contains values for Weight_Lbs that are identical to the values in the previous task.

Conclusion

It is quite likely that you will need to perform one or more of the tasks described in this chapter on a regular basis. Since the character-to-numeric conversion is one of the most common tasks, you may choose to store the conversion macro in your macro library.

Also keep in mind that using the special keywords _NUMERIC_ and _CHARACTER_ to define an array can save you immense time when you need to perform an operation on all character or numeric variables in a data set.

Chapter 2: Grouping Data

Introduction

This chapter describes how to use individual values such as ages or income and create new variables that represents groups. For example, you may want to group ages in 20-year intervals or you may want to convert income into deciles. In this chapter, you will see several ways to accomplish these goals.

Task: Grouping values using IF-THEN-ELSE statements

Keywords

Grouping values

If-Then-Else statement

For this example, you have a variable called Heart_Rate that represents a person's heart rate in beats per minute. You want to create a new variable called HR_Group that represents heart rate ranges. As with all SAS programming, assume that Heart_Rate may possibly be a missing value. Here is the code:

Program 2.1: Using IF-THEN-ELSE logic to group values

```
*Grouping values using if-then-else statements;
data Grouped;
   length HR_Group $ 10.;
   set Blood_Pressure(keep=Subj Heart_Rate);
   if missing(Heart_Rate) then HR_Group = ' ';
   else if Heart_Rate lt 40 then HR_Group = '<40';
   else if Heart_Rate lt 60 then HR_Group = '40-<60';
```

```
    else if Heart_Rate lt 80 then HR_Group = '60-<80';
    else if Heart_Rate lt 100 then HR_Group = '80-<100';
    else HR_Group = '100 +';
run;
```

The first step is to use a LENGTH statement to define the length of the new HR_Group variable. You use the MISSING function to test if Heart_Rate is missing. If so, you set HR_Group to a character missing value. Once you do this, it is straightforward to test the HR_Rate values and create the HR_Group variable. Here are the first 10 observations from data set Grouped:

Listing of data set Grouped

HR_Group	Subj	Heart_Rate
60-<80	1	71
	2	.
60-<80	3	78
60-<80	4	60
60-<80	5	79
80-<100	6	96
40-<60	7	51
	8	.
40-<60	9	54
60-<80	10	65

Task: Grouping values using user-defined formats

Keywords

Grouping data

Formats

PUT function

You can use a format along with a PUT function to accomplish the same goal as the previous example. This technique is quite easy to program—it is also an efficient method for grouping values.

The first step is to define a format with your desired ranges. Then, you can use a PUT function to create the new variable representing your groups. This example places heart rates into the same groups as the previous example:

Program 2.2: Using formats to group variables

```
*Grouping values using formats;
proc format;
    value HRgrp 0 - <40  = '<40'
               40 - <60 = '40-<60'
               60 - <80 = '60-<80'
               80 - <100 = '80-<100'
              100 - high = '100 +';
run;
data Grouped;
    set Blood_Pressure(keep=Subj Heart_Rate);
    HR_Group = put(Heart_Rate,HRgrp.);
run;
```

You name the format HRgrp and use the same ranges as you used previously. Next, you read in values of Heart_Rate and use a PUT function to create HR_Group. You do not need to use a LENGTH statement because SAS uses the longest formatted value for the length of the new variable. Missing values are not a problem either. If Heart_Rate is missing, HR_Group will also be missing. (Because it does not match any of the five ranges you have defined in the format, it remains a missing value—albeit, a character missing value.)

Values for HR_Group are identical to the values in the previous program.

Task: Creating groups using PROC RANK

Keywords

Grouping data

PROC RANK

Deciles

Another way to group data is to place your values into a number of equal groups. This is especially useful for creating statistical models with variables that have highly skewed distributions. For example, you may want to include income in a predictive model. Since income is usually positively skewed (that is, there are some values that are much larger than the average), you may decide to create deciles (10 equal groups) of income.

You can use PROC RANK to convert a continuous variable into any number of equal groups. Before you get started with the task at hand, take a moment to review PROC RANK.

The syntax is:

```
PROC RANK DATA=data-set-name OUT=output-data-set-name
          GROUPS=n;
   VAR list of variables;
   RANKS list of variables holding the rank values;
run;
```

Here is a simple example that demonstrates how PROC RANK works when you include the GROUPS option. Data set Raw_Data contains three variables: Subj, X, and Y, and four observations. Here is a listing:

Subj	X	Y
001	3	10
002	7	20
003	2	30
004	4	40

First, run PROC RANK without the GROUPS= option, like this:

Program 2.3: Demonstrating PROC RANK

```
proc rank data=Raw_Data out=Rank_Data;
   var X;
   ranks Rank_X;
run;
```

You want to create a new data set (Rank_Data) that contains a new variable that you decided to call Rank_X. Values of this variable are the ranks of X. If you supply a list of variables on the VAR statement, be sure to include another list that contains the same number of variables on the RANKS statement. Each variable you list on the RANKS statement will be the rank of the corresponding variable on the VAR statement.

To create ranks, PROC RANK sorts the values of the variables you specify on the VAR statement and gives the smallest value a rank of 1, the next smallest value a rank of 2, and so forth. Here is a listing of data set Rank_Data:

Subj	X	Y	Rank_X
001	3	10	2
002	7	20	4
003	2	30	1
004	4	40	3

There are several things to notice about this output data set. First, Rank_X represents the ranks of variable X. By the way, and this is important, if you leave out the RANKS statement, PROC RANK will replace the original variables (X in this example) with the ranks of these variables. I strongly recommend that whenever you use PROC RANK, include both VAR and RANKS statements.

Next, notice that variable Y is in the output data set even though Y was not listed on the VAR statement. This is a feature of PROC RANK: Every variable in the input data set is included in the output data set. Therefore, if you don't want all of the variables in the input data set to appear in the output data set, include a KEEP= data set option on the input data set, listing the variables you want.

To divide X into two equal groups, use the option GROUPS=2 like this:

Program 2.4: Demonstrating the GROUPS=n option of PROC RANK

```
proc rank data=Raw_Data(keep=Subj X) out=Rank_Data groups=2;
   var X;
   ranks Rank_X;
run;
```

You include a KEEP= data set option so that the output data set will only contain the variables Subj, X, and Rank_X.

Here is the result:

Subj	X	Rank_X
001	3	0
002	7	1
003	2	0
004	4	1

Rank_X now represents groups rather than ranks. When you use the GROUPS=*n* option, PROC RANK numbers the groups starting with zero (why? I have no idea!). If you use GROUPS=10, you will have groups going from zero to nine.

Now that you see how PROC RANK works, you can use it with the Blood_Pressure data set to create quintiles of heart rate, like this:

Program 2.5: Using PROC RANK to create quintiles of Heart_Rate

```
*Grouping using PROC RANK;
proc rank data=Blood_Pressure(Keep=Subj Heart_Rate)
          out=Grouped groups=5;
   var Heart_Rate;
   ranks HR_Group;
run;
```

Here is a listing of the first 12 observations in data set Grouped:

Subj	Heart_Rate	HR_Group
1	71	2
2	.	.
3	78	3
4	60	1
5	79	3
6	96	4
7	51	1
8	.	.
9	54	1
10	65	1
11	55	1
12	48	0

Remember, the group numbers start from zero. In this listing you have five heart rate groups with values from 0 to 4.

PROC RANK with the GROUPS= option is a very easy way to divide any variable into groups.

Conclusion

You can use a variety of methods to create variables that represent groups of continuous variables. Although IF-THEN-ELSE statements in a DATA step are the most common method for grouping values, keep in mind that user-written formats along with a PUT function are an easy and efficient way to create groups. If you want to create a number of groups of equal size, keep PROC RANK in mind.

Chapter 3: Summarizing Data

Introduction

You can use procedures such as PROC MEANS, PROC UNIVARIATE, and PROC TABULATE to print summary information such as means, medians, standard deviations, etc., but there are occasions where you need this summary information in a SAS data set.

You might use this summary information for further analysis or you might need to combine this summary information with the detail information in your original SAS data set.

This chapter describes how to accomplish these tasks.

Task: Using PROC MEANS to create a data set containing summary information

Keywords

PROC MEANS

Summary data set

Grand mean

Let's start out by computing the mean of all observations and outputting this value to a SAS data set. Later in this chapter, you will see how to combine summary data in one data set with detail data in another. This program uses the Blood_Pressure data set and computes the mean for the variable Heart_Rate:

Program 3.1: Computing the mean of all observations and outputting it to a SAS data set

```
*Computing the mean of all observations and outputting it to a
 SAS data set;
proc means data=Blood_Pressure noprint;
    var Heart_Rate;
    output out=Summary(keep=Mean_HR) mean=Mean_HR;
run;
```

The NOPRINT option instructs the program not to produce any printed output because you only want the value in a SAS data set. You may, if you wish, produce both printed output and a data set. You list the variables you want to summarize with a VAR statement. In this example, you only want statistics on Heart_Rate. The OUTPUT statement is an instruction to PROC MEANS that you want to create an output data set. You name the data set following the keyword OUT=. In this example a KEEP= data set option requests that you only want the single variable Mean_HR in the output data set. Without this KEEP=data set option, PROC MEANS adds two other variables in the output data set: _TYPE_ and _FREQ_ (more about these later). Here is a listing of data set Summary:

Mean_HR
69.3448

This small (OK, very small) data set contains a single observation and one variable (the mean Heart_Rate).

Task: Computing the mean of a variable broken down by values of another variable: Using a BY variable

Keywords

PROC MEANS

BY variable

Summary data set

The next task is similar to the previous task except that this time you want to output means for each value of a BY variable. You first need to sort the data set (in this example by Drug) and then run PROC MEANS, including a BY statement. Here is the code:

Program 3.2: Computing the mean for each value of a BY variable

```
*Computing the mean for each value of a BY variable;
proc sort data=Blood_Pressure;
   by Drug;
run;

proc means data=Blood_Pressure noprint;
   by Drug;
   var Heart_Rate;
   output out=Summary mean=Mean_HR;
run;
```

Notice that in this example, the KEEP= data set option was not used. This allows you to see exactly what PROC MEANS places in its output data set. Here is the listing of Summary:

Drug	_TYPE_	_FREQ_	Mean_HR
Drug A	0	20	73.3000
Drug B	0	20	65.3500
Placebo	0	20	69.3889

This data set contains the mean Heart_Rate for each of the three values of Drug.

When you use a BY statement and create an output data set with PROC MEANS, the _TYPE_ variable will always be zero. Later, when you use a CLASS statement, this variable will take on different values and can be quite useful. The variable _FREQ_ tells you the number of observations in each of the three Drug groups (missing and non-missing).

Task: Computing the mean of a variable broken down by values of another variable: Using a CLASS statement

Keywords

PROC MEANS

CLASS statement

Summary data set

You can accomplish the same goal as the previous example by using a CLASS statement instead of a BY statement. The resulting data set, however, is **not** the same. This is important.

It is not necessary to sort your data set when you use a CLASS statement. Here is the code, followed by a listing of the output data set:

Program 3.3: Using PROC MEANS with a CLASS statement to create a summary data set

```
*Using PROC MEANS to create a summary data set;
proc means data=Blood_Pressure noprint;
   class Drug;
   var Heart_Rate;
   output out=Summary mean=Mean_HR;
run;
```

And here is a listing of data set Summary:

Drug	_TYPE_	_FREQ_	Mean_HR
	0	60	69.3448
Drug A	1	20	73.3000
Drug B	1	20	65.3500
Placebo	1	20	69.3889

Notice that this data set contains one more observation than the data set you created using a BY statement. The first observation (_TYPE_ = 0) is the mean of all the non-missing values of Heart_Rate in the Blood_Pressure data set (called the Grand Mean by statisticians). The remaining observations represent the mean Heart_Rate for each value of Drug. Since you probably do not want the grand mean in your output data set, use the NWAY option with PROC MEANS. This option instructs the procedure to output only the means for each combination of the CLASS variable(s).

The next program is identical to this program except that the NWAY option was included:

Program 3.4: Demonstrating the NWAY option of PROC MEANS

```
*Demonstrating the NWAY option;
proc means data=Blood_Pressure noprint nway;
   class Drug;
   var Heart_Rate;
   output out=Summary mean=Mean_HR;
run;
```

Here is the resulting output data set:

Drug	_TYPE_	_FREQ_	Mean_HR
Drug A	1	20	73.3000
Drug B	1	20	65.3500
Placebo	1	20	69.3889

This data set only contains the mean Heart_Rate for each value of Drug—the grand mean is no longer included.

Task: Have PROC MEANS name the variables in the output data set automatically (the AUTONAME option)

Keywords

AUTONAME

PROC MEANS

Summary data set

You can request that PROC MEANS name all the variables in the output data set by using the OUTPUT option called AUTONAME. This option creates variable names in the output data set by using the variable names on the VAR statement and adding a suffix related to the requested statistic. For example, if you have a variable called Heart_Rate and you request a mean, the variable name that PROC MEANS creates is Heart_Rate_Mean. That is, it takes the name of the statistic, adds an underscore, and the name of the requested statistic. The following program requests four statistics (n, mean, min, and max) and uses the AUTONAME option to name the newly created variables:

Program 3.5: Demonstrating the AUTONAME option on the OUTPUT statement in PROC MEANS

```
*Demonstrating the AUTONAME option;
proc means data=Blood_Pressure noprint nway;
   class Drug;
   var Heart_Rate;
   output out=Summary(drop=_type_ _freq_)
      n= mean= min= max= / autoname;
run;
```

Because AUTONAME is an option on the OUTPUT statement, you place the option following a forward slash. Notice that you follow each statistic by an equal sign but you do not include any variable names following the equal sign. This is a very useful option and by using it, you will always have consistent names in your output data sets. Since you are using the NWAY option, you do not need the _TYPE_ variable in the output data set. You may also choose to drop _FREQ_ as shown in this program. Here is a listing of data set Summary:

Drug	Heart_Rate_N	Heart_Rate_Mean	Heart_Rate_Min	Heart_Rate_Max
Drug A	20	73.3000	42	98
Drug B	20	65.3500	37	89
Placebo	18	69.3889	38	116

Task: Creating multiple output data sets from PROC MEANS, each with a different combination of CLASS variables

Keywords

PROC MEANS

CHARTYPE option

Summary data set

There are times when you want summary data sets for various combinations of your CLASS variables. For example, you may want to see the average heart rate for each level of Drug, each level of Gender, and each combination of Drug and Gender. As you saw previously, if you leave off the NWAY option the output data set will contain all of these values, along with _TYPE_.

The interpretation of the values of _TYPE_ is somewhat complicated. To make things easier, use the CHARTYPE procedure option to convert _TYPE_ to a character string of ones and zeros. An explanation of this string is provided following the output. Here is the program:

Program 3.6: Demonstrating the CHARTYPE option with PROC MEANS

```
*Demonstrating the CHARTYPE procedure option;
proc means data=Blood_Pressure noprint chartype;
   class Drug Gender;
   var Heart_Rate;
   output out=Summary mean=Mean_HR;
run;
```

A listing of data set Summary is shown next:

Drug	Gender	_TYPE_	_FREQ_	Mean_HR
		00	58	69.2500
	F	01	28	71.3846
	M	01	30	67.4000
Drug A		10	20	73.3000
Drug B		10	20	65.3500
Placebo		10	18	69.0625
Drug A	F	11	10	81.2000
Drug A	M	11	10	65.4000
Drug B	F	11	10	63.9000
Drug B	M	11	10	66.8000
Placebo	F	11	8	67.5000
Placebo	M	11	10	70.0000

The four values of _TYPE_ are '00', '01', '10', and '11'. Here's how to interpret these values: The two CLASS variables are Drug and Gender. Imagine that you place the first character of _TYPE_ under Drug, and the second character of _TYPE_ under Gender. When you have a '1' under your class variable, the values you see for the mean are for each level of that CLASS variable. When you have a '0' under your class variable, the values you see are for all levels of that CLASS variable, as shown in the following table:

Class	Drug	Gender	Interpretation
	0	0	The grand mean—all Drugs and all Genders
	0	1	Mean heart rate for each value of Gender
	1	0	Mean heart rate for each value of Drug
	1	1	Mean heart rate for each combination of Drug and Gender

For example, the mean Heart_Rate for females (regardless of which drug they are taking) is 71.3846. The mean Heart_Rate for females on drug A is 73.3000.

The next step is to separate the observations in the summary data set into separate data sets for each value of _TYPE_. The following DATA step does this:

Program 3.7: Using the _TYPE_ variable to send summary data to separate data sets

```
data Grand(drop=Drug Gender)
     ByGender(drop=Drug)
     ByDrug(drop=Gender)
     ByDrugGender;
     drop _type_ _freq_;
   set Summary;
   if _type_ = '00' then output Grand;
   else if _type_ = '01' then output ByGender;
   else if _type_ = '10' then output ByDrug;
   else if _type_ = '11' then output ByDrugGender;
run;
```

This single DATA step produces four data sets. Because the value of Drug is missing for all observations in the ByGender data set and the value of Gender is missing for all observations in the ByDrug datadata set, you use a DROP= data set option to remove these variables. Here is a listing of each of these data sets:

Data set Grand

Mean_HR
69.25

Data set ByGender

Gender	Mean_HR
F	71.3846
M	67.4000

Data set ByDrug

Drug	Mean_HR
Drug A	73.3000
Drug B	65.3500
Placebo	69.0625

Data set ByDrugGender

Drug	Gender	Mean_HR
Drug A	F	81.2
Drug A	M	65.4
Drug B	F	63.9
Drug B	M	66.8
Placebo	F	67.5
Placebo	M	70.0

You can now see how useful the CHARTYPE option can be.

Task: Combining summary information (a single mean) with detail data: Using a conditional SET statement

Keywords

Grand mean

Combining summary and detail data

The remaining programs in this chapter show how to combine summary data with detail data. For this first example, you want to add the mean Heart_Rate (Mean_HR) to each observation in the Blood_Pressure data set. There are several ways to accomplish this.

The "classic" approach is to use a conditional SET statement. This is a clever trick that I learned many years ago at a SAS conference (in a talk by Neil Howard). By the way, back then SAS Global Forum was called SUGI (SAS Users Group International).

First the program, then the explanation:

Program 3.8: Adding the grand mean to each observation in your data set

```
*Program to compare each person's heart rate with the mean heart
 rate of all the observations;
proc means data=Blood_Pressure noprint;
   var Heart_Rate;
   output out=Summary(keep=Mean_HR) mean=Mean_HR;
run;
```

You first use PROC MEANS to create the output data set. You use the NOPRINT procedure option since you do not want any printed output, just the summary data set. The Summary data set contains one observation and one variable (Mean_HR). The trick is how to add this single value to each observation in the original Blood_Pressure data set. Many programmers first think of using a MERGE statement. The problem with this solution is that there is no BY variable. You can use a conditional set statement to accomplish your goal as shown in the following short DATA step:

Program 3.9: Demonstrating a conditional SET statement (to combine summary data with detail data)

```
data Percent_of_Mean;
   set Blood_Pressure(keep=Heart_Rate Subj);
   if _n_ = 1 then set Summary;
   Percent = round(100*(Heart_Rate / Mean_HR));
run;
```

It's important to understand exactly what is happening here and why you need to set the summary data set only once. Let's "play computer." In the first iteration of the DATA step, you bring in the first observation from the Blood_Pressure data set (with the two variables Heart_Rate and Subj). Since this is the first iteration of the DATA step, _N_ is equal to 1, the IF statement is evaluated as true, and the SET SUMMARY statement executes, bringing in the one (and only) observation from the Summary data set.

In the second iteration of the DATA step, you first bring in the second observation from the Blood_Pressure data set. Since _N_ is now equal to 2, the IF statement is not true and the second SET statement does not execute. What is the value of Mean_HR in the PDV (program data vector)? You may think it is a missing value. But SAS only sets values being read from raw data or values defined in an assignment statement to missing at the top of the DATA step. The value of Mean_HR comes from a SAS data set—therefore it is automatically retained. That is, the value is not initialized to missing. Since the SET SUMMARY statement executes only once, the value of Mean_HR remains in the PDV and is added to every observation in the Percent_of_Mean data set.

To demonstrate a practical reason why you may want to combine detail and summary data, the program computes a variable called Percent that represents the person's heart rate as a percent of the average of all heart rates.

Here are the first five observations in the Percent_of_Mean data set:

Subj	Heart_Rate	Mean_HR	Percent
1	71	69.3448	102
2	.	69.3448	.
3	78	69.3448	112
4	60	69.3448	87
5	79	69.3448	114

You may ask why do I need the conditional set statement? What happens if I just write:

```
set Summary;
```

The first observation is fine. However, in the second iteration of the DATA step, SAS will try to read the second observation in the Summary data set—and there isn't one. This causes the DATA step to stop and you have a data set with only one observation. In general, when any data set reaches the end of file in a DATA step, the DATA step stops.

Task: Combining summary information (a single mean) with detail data: Using PROC SQL

Keywords

Grand mean

Combining summary and detail data

PROC SQL

Cartesian product

A very simple and elegant solution to combining detail and summary data uses PROC SQL and something called a Cartesian product. A Cartesian product is a combination of every observation in one data set with every observation in another data set. For example, if one data set has 10 observations and the other has 15 observations, the Cartesian product data set will have 10 times 15 or 150 observations. Since the summary data set has only one observation, the Cartesian product data set simply adds this one observation to each observation in the detail data set. Here is the code:

Program 3.10: Combining summary information with detail data using PROC SQL

```
*Solution using PROC SQL;
proc sql;
   create table Percent_of_Mean as
   select Subj,Heart_Rate,Mean_HR
   from Blood_Pressure, Summary;
quit;
```

If you are not familiar with PROC SQL, the CREATE TABLE statement is similar to a DATA statement. You use a SELECT clause to tell PROC SQL what variables you want to select and you use a FROM clause to name the data sets that it will use to create the Cartesian product.

A listing of the first five observations follows:

Subj	Heart_Rate	Mean_HR
1	71	69.3448
2	.	69.3448
3	78	69.3448
4	60	69.3448
5	79	69.3448

You see that the mean Heart_Rate has been added to each observation in the Blood_Pressure data set.

Task: Combining summary information (a single mean) with detail data: Using PROC SQL without using PROC MEANS

Keywords

Grand mean

Combining summary and detail data

PROC SQL

Cartesian product

Mean function

If you are comfortable with PROC SQL, you can use it instead of PROC MEANS to create the mean and add the mean to each observation in the original data set, all in one step. In this example, you include a calculation for Heart_Rate as a percent of the mean, directly in the SELECT clause.

Program 3.11: Combining summary information with detail data using PROC SQL without using PROC MEANS

```
*PROC SQL solution not using PROC MEANS;
proc sql;
   create table Percent_of_Mean as
   select Subj,Heart_Rate, round(100*Heart_Rate / mean(Heart_Rate))
      as Percent
   from Blood_Pressure;
quit;
```

The denominator for this calculation contains the MEAN function. PROC SQL first needs to compute the mean Heart_Rate and then complete the calculation. If you look in the SAS Log, you will notice the message:

```
NOTE: The query requires remerging summary statistics back with the
original data.
```

The resulting output is the same as in the previous task.

This is a very nice use of PROC SQL, producing the desired data set in one step instead of running PROC MEANS and a DATA step.

Task: Combining summary information (a single mean) with detail data: Using a macro variable

Keywords

Grand mean

Combining summary and detail data

Macro variable

CALL SYMPUTX

The next program demonstrates yet one more method of combining detail and summary information. Once you have created the Summary data set (using PROC MEANS), you can run a short data _NULL_ step to assign the mean (Mean_HR) to a macro variable. You can then use this macro variable in any subsequent DATA or PROC step. Submit the following short DATA step to create the macro variable:

Program 3.12: Combining summary information with detail data using a macro variable

```
*Solution using a macro variable;
data _null_;
   set summary;
   call symputx('Macro_Mean',Mean_HR);
run;
```

The two arguments to CALL SYMPUTX are the name of a macro variable and a DATA step variable that contains the value. In this example, you want to assign the value of Mean_HR to the macro variable Macro_Mean.

An even more elegant solution was suggested to me by Mark Jordan, a veteran SAS instructor. By using an INTO clause, you can create a macro variable directly with PROC SQL like this:

Program 3.13: Using PROC SQL to create the macro variable

```
proc sql noprint;
   select mean(Heart_Rate)
   into :Macro_Mean
   from Blood_Pressure;
quit;
```

To demonstrate how you use a macro variable, the short DATA step shown next computes the percent of mean for each subject in the Blood_Pressure data set:

Program 3.14: Demonstrating how to use a macro variable in a DATA step

```
data Percent_of_Mean;
   set Blood_Pressure(keep=Heart_Rate Subj);
   Percent = round(100*(Heart_Rate / &Macro_Mean));
run;
```

The conditional SET statement is, perhaps, the most classic method for combining a single observation with each observation in a detail data set. However, any of the alternate methods demonstrated here are fine. Use the one you are most comfortable with.

Task: Combining summary data with detail data—for each category of a BY variable

Keywords

Combining summary and detail data

BY variables

Merge

When you compute means for each value of another variable (Gender, for example), you can use a simple MERGE statement to combine the summary data with the detail data because you now have a BY variable that you can use. The next example computes the mean Heart_Rate for men and women and uses those means to compute each person's heart rate as a percent of the mean for all subjects of the same gender.

Program 3.15: Creating a summary data set using PROC MEANS with a CLASS statement

```
*Program to compare each person's heart
 rate with the mean heart rate for each
 value of Gender;

proc means data=Blood_Pressure noprint nway;
   class Gender;
   var Heart_Rate;
   output out=By_Gender(keep=Gender Mean_HR) mean=Mean_HR;
run;
```

This example uses a CLASS statement to compute the mean Heart_Rate for each value of Gender. You need to use the NWAY option since you do not want the mean for both genders combined. An alternative would be to use PROC SORT (by Gender) and use a BY statement instead of a CLASS statement.

Here is a listing of the By_Gender data set:

Gender	Mean_HR
F	71.3846
M	67.4000

All you need to do now is to merge the By_Gender data set with the Blood_Pressure data set, using Gender as your BY variable. Of course, you need to sort the Blood_Pressure data set first. (Note: You do not have to sort the By_Gender data set because PROC MEANS sorts it in the order of the CLASS variables.) Here is the code:

Program 3.16: Combining summary data with detail data for each category of a BY variable

```
proc sort data=Blood_Pressure;
   by Gender;
run;

data Percent_of_Mean;
   merge Blood_Pressure(keep=Heart_Rate Gender Subj) By_Gender;
   by Gender;
   Percent = round(100*(Heart_Rate / Mean_HR));
run;

*Put the observations back in Subj order;
proc sort data=Percent_of_Mean;
   by Subj;
run;
```

And, finally, here is a listing of the first five observations of the Percent_of_Mean data set:

Subj	Gender	Heart_Rate	Mean_HR	Percent
1	M	71	67.4000	105
2	F	.	71.3846	.
3	M	78	67.4000	116
4		60	.	.
5	M	79	67.4000	117

The missing value of Percent for subject 2 is due to the missing value for Heart_Rate—the missing value of Percent for subject 4 is due to the missing value for Gender.

Conclusion

In this chapter, you saw how to create summary data sets using PROC MEANS and various ways to combine this summary data with the detail data from your original data set. As with most aspects of SAS programming, there are several different ways to solve a problem. I expect the methods you choose, from the ones discussed in this chapter, are the ones that you are most comfortable with. If you are dealing with very large data sets, you may decide to try several solutions and choose the most efficient one.

Chapter 4: Combining and Updating SAS Data Sets

Introduction

This chapter shows you several ways to concatenate SAS data sets or to add new data to an existing data set. A simple SET statement, listing the data sets you want to concatenate, is one approach. PROC APPEND, under the right circumstances, provides an extremely efficient solution. If your two data sets contain the same variables, but the length of some of the character variables is not consistent in the two data sets, the problem becomes more complicated. This chapter discusses how to accomplish this last task and proceeds to develop a macro to automate the process.

Another common task involves updating a data set with new values. For example, you might have a data set with product identifiers and prices and you wish to update some of the prices. You will see several innovative ways to accomplish this task. You can also use the same techniques to correct data errors in a SAS data set.

Finally, the last task in this chapter discusses how to perform a "fuzzy" match—matching names from two SAS data sets where the names may not be spelled exactly the same.

Task: Concatenating two SAS data sets—Using a SET statement

Keywords

Concatenating data sets

SET statement

Suppose you have two data sets, Name1 and Name2, and you want to add the data from Name2 to the data in Name1. To help visualize the process, here are the listings of the two data sets:

Listing of data set Name1

Name	Gender	Age	Height	Weight
Horvath	F	63	64	130
Chien	M	28	65	122
Hanbicki	F	72	62	240
Morgan	F	71	66	160

Listing of data set Name2

Name	Gender	Age	Height	Weight
Snow	M	51	76	240
Hillary	F	35	69	155

The most straightforward way to put these two data sets together and create a new data set is by using a SET statement. Before putting these two data sets together, it is a good idea to check the storage lengths of all of the character variables. If there are any differences, the length of the variables in the first data set listed on the SET statement will determine the length of the variables in the newly created data set. This can cause truncation of your values if the length of any variable is shorter in the first data set.

Excerpts from PROC CONTENTS for data sets Name1 and Name2 follow:

Data Set Name	WORK.NAME1	Observations	4

Alphabetic List of Variables and Attributes			
#	Variable	Type	Len
3	Age	Num	8
2	Gender	Char	1
4	Height	Num	8
1	Name	Char	10
5	Weight	Num	8

Data Set Name	WORK.NAME2	Observations	2

Alphabetic List of Variables and Attributes			
#	Variable	Type	Len
3	Age	Num	8
2	Gender	Char	1
4	Height	Num	8
1	Name	Char	10
5	Weight	Num	8

The lengths of all the character variables are identical in each of the two files. Later in this chapter, you will see how to approach this task when the lengths are not the same.

You can now go ahead and concatenate the two files using a SET statement as follows:

Program 4.1: Concatenating SAS data sets using a SET statement

```
*Concatenating SAS data sets;
data Combined;
   set Name1 Name2;
run;
```

Data set Combined contains six observations, four from Name1 and two from Name2.

Task: Concatenating two SAS data sets—Using PROC APPEND

Keywords

Concatenating data sets

PROC APPEND

Program 4.1 requires SAS to read all the observations from Name1 as well as Name2. If the first data set (Name1 in the example) is large, this uses considerable computer resources. An alternative is to use PROC APPEND to add the observations from Name2 to the end of data set Name1. PROC APPEND jumps directly to the end of Name1 and starts adding the observations from Name2. In this case, since the two data sets have identical variables and attributes, this is a very efficient solution.

Because PROC APPEND adds the new data to the first data set (and thereby changes it), you can't undo the process. You might consider making a copy of the first data set before running PROC APPEND.

Program 4.2 demonstrates how you can use PROC APPEND to concatenate two data sets:

Program 4.2: Concatenating SAS data sets using PROC APPEND

```
*Method 2 - Using PROC APPEND;
proc append base=Name1 data=Name2;
run;
```

You name the first data set with the BASE= option, and the data to be added with the DATA= option. After you run this program, the Name1 data contains six observations.

Task: Concatenating two SAS data sets with character variables of different lengths

Keywords

Concatenating SAS data sets

PROC APPEND

FORCE option

To demonstrate what happens if the two data sets being concatenated have character variables of different lengths, a new data set, Name3, was created. A listing of the data set and output from PROC CONTENTS showing the variable attributes follows:

Listing of data set Name3

Name	Gender	Age	Height	Weight
Zemlachenko	M	55	72	220
Reardon	M	27	75	180

File attributes from Name3

Data Set Name	WORK.NAME3	Observations	2

Alphabetic List of Variables and Attributes			
#	Variable	Type	Len
3	Age	Num	8
2	Gender	Char	2
4	Height	Num	8
1	Name	Char	18
5	Weight	Num	8

The length of Name in data sets Name1 and Name2 is 10. In data set Name3 it is 18; the length of Gender in data sets Name1 and Name2 is 1, and the length of Gender in data set Name3 is 2. If you attempt to run PROC APPEND using data sets Name2 and Name3, you will see the following message in the SAS Log:

```
WARNING: Variable Name has different lengths on BASE and DATA files
         (BASE 10 DATA 18).
WARNING: Variable Gender has different lengths on BASE and DATA files
         (BASE 1 DATA 2).
ERROR: No appending done because of anomalies listed above.
       Use FORCE option to append these files.
NOTE: 0 observations added.
```

Since the length of Name in data set Name2 is shorter than the length of Name in data set Name3, names from the Name3 data set may be truncated. Even though the length of Gender is longer in Name3, all of the Gender values are single letters. To tell SAS to go ahead and run PROC APPEND even though there are different lengths in the two files, you need to use the FORCE procedure option like this:

Program 4.3: Running PROC APPEND with the FORCE option

```
*Attempting to combine data sets with character variables
 of different lengths using the FORCE option;
proc append base=Name2 data=Name3 force;
run;
```

PROC APPEND now runs, concatenating the two data sets. The storage lengths of the character variables are derived from the BASE= data set (Name2 in this example). The portion of PROC CONTENTS that shows the variable attributes in the concatenated data set follows:

Data Set Name	WORK.NAME2	Observations	6

Alphabetic List of Variables and Attributes			
#	Variable	Type	Len
3	Age	Num	8
2	Gender	Char	1
4	Height	Num	8
1	Name	Char	10
5	Weight	Num	8

Notice that Name has a length of 10 and Gender has a length of 1.

Task: Concatenating two SAS data sets that contain character variables of different lengths and controlling the length of the character variables

Keywords

Concatenating SAS data sets

SET statement

LENGTH statement

What do you do when the two data sets you want to concatenate have character variables of different lengths? If all the lengths in the first data set are longer than those in the second data set, you can use a simple SET statement listing the two data sets in order, or you can use PROC APPEND with the FORCE option. Of course, life is never that simple. Applying Murphy's Law, you can be sure that there will be some lengths that are longer in the first data set and some that are longer in the second data set.

The next program shows how to use a LENGTH statement before the SET statement to assign a length for selected character variables. Here is the program:

Program 4.4: Using a LENGTH statement before the SET statement to assign appropriate lengths

```
Data Combined;
   length Gender $2 Name $ 18;
   set Name2 Name3;
run;
```

Because the LENTH statement precedes the SET statement, the length of Name is 18 and the length of Gender is 2.

After running Program 4.4, the variable Name in the newly created data set (Combined) has a length of 18 and Gender has a length of 2.

Task: Developing a macro to concatenate two SAS data sets that contain character variables of different lengths

Keywords

Concatenating SAS data sets

Concatenating macro

The macro developed in this task concatenates two data sets and automatically assigns the appropriate length for each character variable.

This macro uses an interesting technique: It writes out a SAS program to an external file and uses a %INCLUDE statement to execute the program. Here is a summary of the approach taken:

- Use PROC CONTENTS to output two data sets containing the lengths of all the character variables in the data sets you want to concatenate.
- Compare the lengths from the two data sets and choose the longer length for each character variable.
- Write a LENGTH statement using these lengths.
- Using PUT statements, write a short DATA step using these lengths.
- Run this program using a %INCLUDE statement.

To automate the process, the first step is to determine the lengths of all the character variables in both data sets using PROC CONTENTS:

Program 4.5: Using PROC CONTENTS to output a data set of character variable storage lengths

```
proc contents data=Name2 noprint
   out=Out1(keep=Name Type Length where=(Type=2));
run;
proc contents data=Name3 noprint
   out=Out2(keep=Name Type Length where=(Type=2));
run;
```

You use an OUTPUT statement to output the variable attributes to a SAS data set. The variables of interest are Name, Type, and Length (PROC CONTENTS chose those names). The variable Type has a value of 1 for numeric variables and 2 for character variables. The WHERE= data set option selects information only on the character variables (Type = 2).

To help visualize the process, here is a listing of data sets Out1 and Out2:

Listing of data set Out1

NAME	TYPE	LENGTH
Gender	2	1
Name	2	10

Listing of data set Out2

NAME	TYPE	LENGTH
Gender	2	2
Name	2	18

The next step is to combine the two data sets created by PROC CONENTS into a single data set. You do this with a MERGE statement, using a RENAME= option on the second data set (renaming Length to Length2) and using Name as the BY variable.

Here is a listing of the merged data set:

NAME	TYPE	LENGTH	Length2
Gender	2	1	2
Name	2	10	18

A quick side-note concerning the MERGE statement: You can set a system option called MERGENOBY equal to either NOWARN, WARN, or ERROR. If you choose ERROR and you forget a BY statement after a MERGE, the DATA step will stop and you will get an error message in the SAS Log. If you set this option to WARN, the program will perform the merge and a warning message will be printed in the Log. The default value for the MERGENOBY option is NOWARN. With NOWARN set, the merge will proceed and it will merge the first observation in file one with the first observation in file two, and so forth. It is very unlikely that this is something you want to do, so I recommend that you set this option to WARN or ERROR. I have mine set to ERROR.

The next steps are 1) Write a DATA statement to the external file. 2) Compare the two values, Length and Length2, and determine which is larger. 3) Write a length statement for each character variable using this length. 4) Write out a SET and RUN statement. 5) Use a %INCLUDE statement to run the program created in the external file.

To accomplish these goals, run the following DATA step:

Program 4.6: Using a DATA step to write out a SAS program to an external file

```
data _null_;
    merge Out1
          Out2(rename=(Length=Length2))
          end=last;
    by Name;
    file "c:\books\tasks\Combined.sas";

/* Step 1 */
    if _n_ = 1 then put "Data Combined;";

/* Step 2 */
    L = max(Length,Length2);

/* Step 3 */
    put "   length " Name " $ " L 3. ";";

/* Step 4 */
    if Last then do;
        put "   set Name1 Name2;";
        put "run;";
    end;
run;

/* Step 5 */
%include "combined.sas";
```

When _N_ is equal to 1, you write out the DATA statement. Then for each character variable, you write out a LENGTH statement using the longer length.

The MERGE statement included an END= option. You need this so that you can write out the final lines of the program (a SET and RUN statement) after all the LENGTH statements have been written.

To help clarify this somewhat confusing process, here is a listing of the external text file you named Combined.sas:

Program 4.7: Listing of the SAS program created by Program 4.6

```
Data Combined;
   length Gender  $   2;
   length Name  $  18;
   set Name1 Name2;
run;
```

Converting this program to a macro is straightforward. You need to substitute macro variables for the two input SAS data sets and the concatenated data set. So finally, here is the macro:

Program 4.8: A macro to concatenate two SAS data sets with character variables of different lengths

```
%macro Concatenate
    (Dsn1=,       /*Name of the first data set    */
     Dsn2=,       /*Name of the second data set   */
     Out=         /*Name of combined data set     */);

   proc contents data=&Dsn1 noprint
      out=out1(keep=Name Type Length where=(Type=2));
   run;

   proc contents data=&Dsn2 noprint
      out=out2(keep=Name Type Length where=(Type=2));
   run;

   data out1;
      set out1;
      Name = propcase(Name);
   run;

   data out2;
      set out2;
      Name = propcase(Name);
   run;

   data _null_;
      file "c:\books\tasks\combined.sas";
      merge out1 out2(rename=(Length=Length2)) end=Last;
      by Name;
      if _n_ = 1 then put "Data &out;";
      l = max(Length,Length2);
      put "   length " Name " $ " L 2. ";";
      if Last then do;
         put "   set &Dsn1 &Dsn2;";
         put "run;";
      end;
   run;

   %include "c:\books\tasks\Combined.sas";

%mend concatenate;
```

This macro takes the extra step of converting all of the variable names to proper case, because if they were not the same case in each data set, the MERGE would not work.

Let's test the macro using the two data sets Name2 and Name3:

Program 4.9: Testing the Concatenate macro

```
*Testing the macro;
%Concatenate(Dsn1=Name2, Dsn2=name3, out=Combined)

title "Contents of COMBINED";
proc contents data=combined;
run;
```

The section of output from PROC CONTENTS shown next demonstrates that the macro worked as desired. Name has a length of 18 and Gender has a length of 2.

Data Set Name	WORK.COMBINED	Observations	4

Alphabetic List of Variables and Attributes			
#	Variable	Type	Len
3	Age	Num	8
1	Gender	Char	2
4	Height	Num	8
2	Name	Char	18
5	Weight	Num	8

You may want to place a copy of this macro in your macro library for solving this common task.

Task: Updating a SAS data set using a transaction data set

Keywords

UPDATE

Transaction data set

For this example, you have a master file (Hardware) that contains item numbers (Item_Number) and prices (Price). A listing of this data set is shown next:

Item_Number	Description	Price
1238	Cross Cut Saw 18 inch	$18.75
1122	Cross Cut Saw 24 inch	$23.95
2001	Nails, 10 penny	$5.57
2002	Nails, 8 penny	$4.59
3003	Pliers, Needle nose	$12.98
3035	Pliers, cutting	$15.99
4005	Hammer, 6 pound	$12.98
4006	Hammer, 8 pound	$15.98
4007	Hammer, sledge	$19.98

You want to change the prices for item numbers 2002 and 4006. You start by creating a data set with the items numbers and the new prices.

Program 4.10: Updating a master file with data from a transaction file

```
*Updating values in a SAS data set using a transaction data set;
Data New_Prices;
   input Item_Number : $4. Price;
datalines;
2002 5.98
4006 16.98
;
```

Here is a listing of the New_Prices data set:

Item_Number	Price
2002	5.98
4006	16.98

The next step is to sort the Hardware and New_Prices data set by Item_Number, as follows:

```
proc sort data=Hardware;
   by Item_Number;
run;

proc sort data=New_Prices;
   by Item_Number;
run;
```

You can either update the original data set or create a new one. It is a good idea to use a different name for the new, updated data. Before showing you the next section of the program, let's take a moment to review the UPDATE statement.

Unlike a MERGE, missing values in the right-most data set (New_Prices in this example) do not replace values in the left-most data set (Hardware in this example). Also, unlike a MERGE, you must follow an UPDATE statement with a BY statement.

The remaining portion of the program, demonstrating the UPDATE statement, follows:

```
Data Hardware_June2012;
   update Hardware New_Prices;
   by Item_Number;
run;
```

The following listing shows that the prices for the two items (2002 and 4006) have been updated:

Listing of data set Hardware_June2012

Item_Number	Description	Price
1122	Cross Cut Saw 24 inch	$23.95
1238	Cross Cut Saw 18 inch	$18.75
2001	Nails, 10 penny	$5.57
2002	Nails, 8 penny	$5.98
3003	Pliers, Needle nose	$12.98
3035	Pliers, cutting	$15.99
4005	Hammer, 6 pound	$12.98
4006	Hammer, 8 pound	$16.98
4007	Hammer, sledge	$19.98

Task: Using a MODIFY statement to update a master file from a transaction file

Keywords
Modify

Transaction file

If you prefer to update your original SAS data set with data from a transaction file, you can use a MODIFY statement. Unlike UPDATE, this technique does not create a new data set but modifies the original data set.

Although you do not have to sort your master file and transaction file when using a MODIFY statement followed by a BY statement, doing so will improve efficiency.

The following short program shows how you can update a master file from data in a transaction file using MODIFY:

Program 4.11: Using a MODIFY statement to update a master file from a transaction file

```
data Hardware;
   modify Hardware New_Prices;
run;
```

After you run this program, the SAS data set Hardware will be identical to the Hardware_June2012 data set created in the previous task.

Task: Updating several variables using a transaction file created with an INPUT method called named input

Keywords
Data cleaning

Updating

Named input

If you need to change the values of several variables, creating the transaction file can become a problem. For example, suppose you want to change the values for Score, Weight, Heart_Rate, DOB, Gender, and Party for several subjects in the Demographic data set, which follows:

Subj	Score	Weight	Heart_Rate	DOB	Gender	Party
1	70	999	76	04NOV1955	Male	NA
2	26	160	62	08APR1955	NA	NA
3	71	195	71	20JUL1955	male	na
4	40	132	74	08JAN1955	Male	Republican
5	999	181	62	15AUG1951	Female	Democrat
6	62	71	52	24JAN1950	Male	democrat
7	24	136	72	26NOV1950	Female	democrat
8	5	174	71	08NOV1950	Female	democrat
9	5	172	47	28DEC1951	Male	Democrat
10	94	173	999	06MAY1953	Male	republican
11	99	170	63	27FEB1950	na	NA
12	10	133	63	18MAR1954	Male	democrat
13	6	131	60	26MAR1951	Female	republican
14	999	140	79	01OCT1950	NA	na
15	999	124	999	12OCT1950	NA	na
16	44	194	72	31DEC1952	Female	republican
17	62	196	68	09MAR1951	Female	democrat
18	57	133	72	15SEP1951	Female	Democrat
19	45	137	86	16NOV1951	NA	Republican
20	90	170	80	01OCT1951	Female	Republican

Using any of the standard INPUT methods, such as list input, column input, or formatted input, is not convenient. If you only need to enter non-missing values for a few variables, you can use an input method called *named input*. Here's how it works.

You use an INFORMAT (or LENGTH) statement to define the length for all of your character variables and an appropriate date informat for any date values you have. Next, you enter each of the variables followed by an equal sign on the INPUT statement. Finally, when you enter data, you use the form: Variable_name = Value.

For this example, you want to change Score to 72 and Party to "Republican" for subject 2, and DOB=26Nov1951 and Weight=140 for subject 7.

Here is the program using named input:

Program 4.12: Using named input to create a transaction data set

```
*Use "Named Input" method to create the transaction data set;
data New_Values;
   informat Gender $6. Party $10. DOB Date9.;
   input Subj= Score= Weight= Heart_Rate= DOB= Gender= Party=;
   format DOB date9.;
datalines;
Subj=2 Score=72 Party=Republican
Subj=7 DOB=26Nov1951 Weight=140
;
```

This may look like a lot of work, but by using this form of input, you do not have to count columns. By the way, you can enter data with named input in any order.

The data New_Values is listed next:

Gender	Party	DOB	Subj	Score	Weight	Heart_Rate
	Republican	.	2	72	.	.
		26NOV1951	7	.	140	.

Notice that any variable not given a value is set to missing. You are now ready to use UPDATE (or MODIFY) to change values in the Demographic data set.

Program 4.13: Updating a master file using a transaction file created with named input

```
proc sort data=Demographic;
   by Subj;
run;
proc sort data=New_Values;
   by Subj;
run;

Data Demographic_June2012;
   update Demographic New_Values;
   by Subj;
run;
```

The newly created data set, Demographic_June2012, is identical to the old Demographic data set except for the new values for subjects 2 and 7.

Task: Matching names from two SAS data sets where the names may not be spelled the same (fuzzy merge)

Keywords

Fuzzy merge

Inexact matching

SPEDIS function

If you are called upon to merge data from two SAS data sets when you do not have a unique identifier in both files, you sometimes have to resort to using names (and other variables) to determine which observations from the two data sets belong together.

Two useful tools for performing this task are the SPEDIS function (stands for spelling distance) that helps you determine words that have similar spelling but are not an exact match, and PROC SQL, that can perform a Cartesian product.

The SPEDIS function takes two character arguments. If these two arguments are identical, the function returns a zero. For each category of spelling error, the function assigns penalty points. Some spelling errors are considered more serious than others and are assigned more penalty points. For example, if the first letter is not the same in the two strings, the function assigns a large penalty. Adding a letter or changing the position of two letters (ie versus ei for example), the function assigns a relatively small penalty.

After the function adds up all the penalty points, it then divides the total penalty points by the length of the first argument to the function. This makes sense, since getting a letter wrong in a three-letter word is a much more serious error than getting a letter wrong in a 10-letter word.

It is a matter of trial and error to decide what you consider a large spelling error. If you try to match two names and allow for a large spelling distance, you may match two names that are actually different—if you insist on a small spelling error before you consider two names to be a match, you may fail to match two names that actually belong to the same person.

PROC SQL can create a Cartesian product between two data sets. This is accomplished by creating every combination of observations in one data set with every observation in a second data set.

To help visualize this, two SAS data sets, Name_One and Name_Two, are listed next:

Listing of Name_One

Name1	DOB1	Gender1
Friedman	14JUL1946	M
Chien	21OCT1965	F
MacDonald	12FEB2001	M
Fitzgerald	04AUG1966	M
GREGORY	05FEB1955	F

Listing of Name_Two

Name2	DOB2	Gender2
Freidman	14JUL1946	M
Chen	21OCT1965	F
McDonald	12FEB2001	M
Fitzgerald	04AUG1966	M
Gregory	05FEB1955	F

The first 13 observations in a Cartesian product of these two data sets are shown next:

Partial listing of the Cartesian product of Name_One and Name_Two

Name1	DOB1	Gender1	Name2	DOB2	Gender2
Friedman	14JUL1946	M	Freidman	14JUL1946	M
Friedman	14JUL1946	M	Chen	21OCT1965	F
Friedman	14JUL1946	M	McDonald	12FEB2001	M
Friedman	14JUL1946	M	Fitzgerald	04AUG1966	M
Friedman	14JUL1946	M	Gregory	05FEB1955	F
Chien	21OCT1965	F	Freidman	14JUL1946	M
Chien	21OCT1965	F	Chen	21OCT1965	F
Chien	21OCT1965	F	McDonald	12FEB2001	M
Chien	21OCT1965	F	Fitzgerald	04AUG1966	M

Name1	DOB1	Gender1	Name2	DOB2	Gender2
Chien	21OCT1965	F	Gregory	05FEB1955	F
MacDonald	12FEB2001	M	Freidman	14JUL1946	M
MacDonald	12FEB2001	M	Chen	21OCT1965	F
MacDonald	12FEB2001	M	McDonald	12FEB2001	M

For this example, you want to find observations in the Cartesian product data set where the date of birth and the gender are the same in each data set and where the spelling distance between the two names is less than a predetermined value. Here is the PROC SQL code to produce a list of possible matches and exact matches:

Program 4.14: Performing a "fuzzy" match between two SAS data sets

```
proc sql;
   create table Possible_Matches as
   select * from Name_One, Name_Two
   where spedis(upcase(Name1),upcase(Name2)) between 1 and 25 and
   DOB1 eq DOB2 and
   Gender1 eq Gender2;
quit;

proc sql;
   create table Exact_Matches as
   select * from Name_One, Name_Two
   where spedis(upcase(Name1),upcase(Name2)) eq 0 and
   DOB1 eq DOB2 and
   Gender1 eq Gender2;
quit;
```

The first PROC SQL step creates a Cartesian product between Name_One and Name_Two. The WHERE clause selects all observations in the Cartesian product data set where the two names are within a spelling distance of 25 (but not zero, which would be an exact match). To be sure to match names that are not in the same case, you use the UPCASE function on both names. As a further effort to determine which observations belong together, you can check that the date of birth and the gender are the same. This program ignores missing values for date of birth and gender. You may want to insist that these variables are not missing before attempting a match.

The second step creates a data set of exact matches. A listing of both data sets follows:

Listing of data set Possible_Matches

Name1	DOB1	Gender1	Name2	DOB2	Gender2
Friedman	14JUL1946	M	Freidman	14JUL1946	M
Chien	21OCT1965	F	Chen	21OCT1965	F
MacDonald	12FEB2001	M	McDonald	12FEB2001	M

Listing of data set Exact_Matches

Name1	DOB1	Gender1	Name2	DOB2	Gender2
Fitzgerald	04AUG1966	M	Fitzgerald	04AUG1966	M
GREGORY	05FEB1955	F	Gregory	05FEB1955	F

Inspection of the Possible_Matches data set shows matches between names that are spelled slightly differently but probably do belong together.

Conclusion

At some time in your programming career, you will need to accomplish most of the tasks described in this chapter. Keep in mind that PROC APPEND can be much more efficient than using a SET statement. When you want to concatenate two data sets that contain character variables of different length, try out the Concatenate macro. Finally, consider using named input when you want to create a transaction file that only contains a few non-missing values out of a large number of variables.

Chapter 5: Creating Formats from SAS Data Sets

Introduction

The basic approach to creating a user-defined format or informat is to run PROC FORMAT and supply labels for each value or a range of values that you want to format. There are times when this becomes very tedious. If the information you need to create a format is already stored in a SAS data set, you can convert this data set into a control data set that can then be used to create your format. This chapter demonstrates how to create a control data set and how to add or modify values in an existing format.

Task: Using a SAS data set to create a format (by creating a control data set)

Keywords

PROC FORMAT

Control data set

Data set Codes contains disease codes and descriptions of these codes. Here is a listing of the Codes data set:

ICD9	Description
020	Plague
022	Anthrax
390	Rheumatic fever

ICD9	Description
410	Myocardial infarction
493	Asthma
540	Appendicitis

The variable ICD9 (stands for International Classification of Diseases version 9) is a code for each of the medical conditions stored in the variable Description. You would like to create a format for each of the ICD9 codes using the value of Description as the format label.

The first step is to create a control data set from the Codes data set. Variables in a control data set are as follows:

Variable Name	Description
Fmtname	A character variable that contains the name of the format you want to create. If you want to create a character format, this value should begin with a dollar sign ($).
Type	Values of Type are 'C' for a character format and 'N' for a numeric format. If the Fmtname starts with a $, the Type variable is optional.
Start	This is a single value or the starting value in a range.
End	If you specify a range, End is the ending value.
Label	This character variable holds the format label.
HLO	This variable allows you to specify values for the keywords High, Low, or Other.

For this task, you want the variable ICD9 to be renamed Start, and Description to be renamed Label. If you want the format to be called $ICDFMT, assign this value to Fmtname. You will see how to use the HLO variable in the next task.

The following DATA step creates a control data set from Codes:

Program 5.1: Creating a control data set from a data set containing codes and labels

```
*Program to create a control data set from a SAS data set;
data Control;
   set Codes(rename=
             (ICD9 = Start
              Description = Label));
   retain Fmtname '$ICDFMT'
          Type 'C';
run;
```

You use the RENAME= data set option to rename the two variables: ICD9 and Description. The value of Fmtname is assigned using a RETAIN statement. This is more efficient than using an assignment statement because an assignment statement would execute for each iteration of the DATA step and the RETAIN statement executes only once. You also set Type equal to 'C' (stands for character). This is not necessary since the format name begins with a dollar sign, but you may want to include it anyway.

Here is a listing of data set Control:

Start	Label	Fmtname	Type
020	Plague	$ICDFMT	C
022	Anthrax	$ICDFMT	C
390	Rheumatic fever	$ICDFMT	C
410	Myocardial infarction	$ICDFMT	C
493	Asthma	$ICDFMT	C
540	Appendicitis	$ICDFMT	C

The next step is to run PROC FORMAT with the Control data set specified with the CNTLIN= option, like this:

Program 5.2: Using a control data set to create a format

```
proc format cntlin=Control;
    select $ICDFMT;
run;

*Using the CNTLIN= created data set;
data disease;
    input ICD9 : $5. @@;
datalines;
020 410 500 493
;
title "Listing of DISEASE";
proc report data=disease nowd headline;
    columns ICD9=Unformatted ICD9;
    define ICD9 / "Formatted Value" width=11 format= $ICDFMT.;
    define Unformatted / "Original Unformatted Value" width=11;
run;
```

You run PROC FORMAT with the option CNTLIN=Control. The SELECT statement in this program is not necessary to create the format—it is a request to list the contents of the format for documentation purposes. It produced the following output:

```
         FORMAT NAME: $ICDFMT  LENGTH:   21   NUMBER OF VALUES:   6
     MIN LENGTH:   1  MAX LENGTH:  40  DEFAULT LENGTH:  21  FUZZ:       0

START               END                LABEL   (VER. V7|V8    19MAR2012:09:07:25)

020                 020                Plague
022                 022                Anthrax
390                 390                Rheumatic fever
410                 410                Myocardial infarction
493                 493                Asthma
540                 540                Appendicitis
```

Finally, the short DATA step and PROC REPORT portion of the program demonstrate how the $ICDFMT format works. PROC REPORT was used instead of PROC PRINT so that you can see both the unformatted and formatted values for the variable ICD9. Here is the listing:

Original Unformatted Value	Formatted Value
020	Plague
410	Myocardial infarction
500	500
493	Asthma

Three of the ICD9 codes are formatted correctly; however, the ICD9 code of 500 does not have a corresponding label and does not get formatted. To add an OTHER category to your format, you set the value of HLO to 'O' (other) and provide the appropriate label for the other category like this:

Program 5.3: Adding an OTHER category to your format

```
*Adding an OTHER category to your format;
data control;
   set Codes(rename=
             (ICD9 = Start
              Description = Label))
              End = Last;
   retain Fmtname '$ICDFMT'
          Type 'C';
   output;
   if Last then do;
      HLO = 'o';
      Label = 'Not Found';
      output;
   end;
run;

title "Adding OTHER Category";
proc format cntlin=Control;
   select $ICDFMT;
run;
```

You need to add two things to your program to create a label for the OTHER category. First, you add the END= option on the SET statement. In this program, the variable Last will have a value of True when you are reading the last observation from data set Codes, and a value of False otherwise. Next, you can use this variable to execute the assignment statements at the end of the DATA step. You set HLO to 'O' and Label to 'Not Found'. You then run PROC FORMAT using the control data set with the OTHER category included as shown here:

```
       FORMAT NAME: $ICDFMT  LENGTH:   21    NUMBER OF VALUES:    7
   MIN LENGTH:   1  MAX LENGTH:  40  DEFAULT LENGTH:  21  FUZZ:        0
```

START	END	LABEL (VER. V7|V8 19MAR2012:09:24:26)
020	020	Plague
022	022	Anthrax
390	390	Rheumatic fever
410	410	Myocardial infarction
493	493	Asthma
540	540	Appendicitis
OTHER	**OTHER**	Not Found

Notice that the new format contains a label for OTHER.

Task: Adding new format values to an existing format

Keywords

Adding formats

CNTLOUT= option

Suppose you want to add some formats to the $ICDFMT format you created previously. One way is to run PROC FORMAT with the option CNTLOUT= to write out the existing Control data set, and use this data set to create a new Control data set containing the additional codes and labels.

As an example, suppose you want to add formats for codes 427.5 (Cardiac Arrest) and 466 (Bronchitis) to the $ICDFMT format. You need to be careful if any of your new labels are longer than any of the existing labels. If that is the case, be sure to use a LENGTH statement to define the new length for the variable Label. Here is the program:

Program 5.4: Adding new formats to an existing format using a CNTLOUT data set

```
*Adding new formats to an existing format using a CNTLOUT data set;
proc format cntlout=Control_Out;
   select $icdfmt;
run;

data New_control;
   length Label $ 25;
   set Control_Out(drop=End) end=Last;
   output;
   if Last then do;
      Hlo = ' ';
      Start = '427.5';
      Label = 'Cardiac Arrest';
      output;
      Start = '466';
      Label = 'Bronchitis - nonspecific';
      output;
   end;
run;

proc format cntlin=New_control;
   select $ICDFMT;
run;
```

You start by running PROC FORMAT with the option CNTLOUT= to output your existing control data set (in this example, you name this data set Control_Out). Since the existing format does not contain ranges, you do not need (or want) the End variable from the Control_Out data set, so you drop it. Next, you use the END= SET option of the SET statement to add the new codes and labels. Notice that the LENGTH statement to define the length for the variable Label comes before the SET statement (otherwise, it would have no effect). Also, you need to set HLO to a missing value because you already assigned a label for the Other category. Finally, you run PROC FORMAT using the new control data set on the CNTLIN= option. A listing of the new format (as a result of the SELECT statement) is shown next:

```
        FORMAT NAME: $ICDFMT  LENGTH:   24   NUMBER OF VALUES:    9
    MIN LENGTH:   1  MAX LENGTH:  40  DEFAULT LENGTH:  21  FUZZ:          0

START              END                LABEL   (VER. V7|V8    19MAR2012:09:36:19)

020                020                Plague
022                022                Anthrax
390                390                Rheumatic fever
410                410                Myocardial infarction
427.5              427.5              Cardiac Arrest
466                466                Bronchitis - nonspecific
493                493                Asthma
540                540                Appendicitis
**OTHER**          **OTHER**          Not Found
```

The two new categories have been successfully added.

Conclusion

There are two situations where it is very convenient to create a format from a SAS data set. One is when you already have a SAS data set containing codes and labels (and it is relatively large). The other is when you want to use formats for table lookup and you have the lookup values in an existing SAS data set. In both of these cases, control data set will save you time.

Chapter 6: Table Lookup Techniques

Introduction

Table lookup is the process where you use one or more key values, such as an item number, and associate it with another value, such as the item's price. SAS has a variety of ways of performing one-way or multi-way table lookup. One easy and straightforward way is to use a MERGE statement. Other, more efficient methods, such as temporary arrays, formats, or informats, store the table information in memory. You will see examples of all of these methods in this chapter.

Task: Performing a one-way table lookup using a MERGE statement

Keywords

Table lookup

MERGE

You have sales data, arranged by year, in your Sales data set and yearly goal data, also arranged by year, in your Goals data set. The first five observations from each of these data sets are listed here:

Listing of data set Sales

Sales_ID	Year	Sales
1234	2004	20.5
1234	2005	22.0
1234	2006	26.0
1234	2007	27.0
1234	2008	37.0

Listing of data set Goals

Year	Goal
2004	20
2005	21
2006	24
2007	28
2008	34

You can use a MERGE statement to add the Goal values to each observation in the Sales data set like this:

Program 6.1: Using a MERGE to perform a table lookup

```
proc sort data=Goals;
   by Year;
run;

proc sort data=Sales;
   by Year;
run;

data Sales_Goals;
   merge Goals Sales;
   by Year;
   Difference = Sales - Goal;
run;

proc sort data=Sales_Goals;
   by Sales_ID Year;
run;
```

First, you need to sort each data set by Year (if it is not already in the correct order). Next, you perform a MERGE on the two data sets, using Year as the BY variable. In this example, you want to see the difference between actual sales and goals for each year for each sales person. The final Sort places the observations in Sales_ID order. The resulting data set is shown next:

Year	Goal	Sales_ID	Sales	Difference
2004	20	1234	20.5	0.5
2005	21	1234	22.0	1.0
2006	24	1234	26.0	2.0
2007	28	1234	27.0	-1.0
2008	34	1234	37.0	3.0

You can improve the efficiency of this operation by using temporary arrays or formats (or informats). This is the topic of the next two tasks.

Task: Performing a one-way table lookup using user-defined informats

Keywords

Table lookup

Informats

Input function

User-defined formats and informats provide you with an extremely powerful tool for performing table lookup. One advantage of this technique is that formats and informats are stored in memory, making the task fast and efficient. The reason you might prefer an informat over a format is that you can use an informat in an INPUT function to create a numeric variable. You will see an example of this in the next program. If you use a format and a PUT function, you will create a character variable.

This first example shows how to create an informat and use that informat to perform the lookup. Following this task, you will see how to use a SAS data set to create a control data set that, in turn, creates the informat.

You create an informat much the same way you create a format. Instead of using a VALUE statement, you use an INVALUE statement. The informat called Goalfmt is created in the short program shown here:

Program 6.2: Creating a user-defined informat to be used for table lookup

```
*Creating the INFORMAT "manually" using PROC FORMAT;
proc format;
   invalue Goalfmt 2004=20
                   2005=21
                   2006=24
                   2007=28
                   2008=34
                   2009=40
                   2010=49
                   2011=60
                   2012=75;
run;
```

Notice how this code resembles the creation of a format. There are two differences: You use the keyword INVALUE instead of VALUE, and you do not place the values to the right of the equal signs in quotes. Goalfmt is a numeric informat, and the result of using it in an INPUT statement or INPUT function is a numeric value. The next step is to use this informat in a DATA step to perform your lookup:

Program 6.3: Using the informat to perform the table lookup

```
data Sales_Goals;
   set Sales;
   Goal = input(put(Year,4.),goalfmt.);
   Difference = Sales - Goal;
run;
```

The INPUT function takes a SAS value as the first argument and "reads" it using the INFORMAT listed as the second argument. You need the PUT function in this example because the first argument to the INPUT function must be a character value. (You could leave it out, and SAS will perform a character-to-numeric conversion automatically and it will print a numeric to character message in the Log. However, this is not recommended.)

Suppose the first year in the Sales data set is 2008. The Goalfmt INFORMAT converts this value to the numeric value 34, and assigns this value to the variable Goal. Because creating an informat (or format) can be tedious, the next section describes how to use a control data set to create informats.

Task: Creating an INFORMAT using a control data set

Keywords

Control data set

Informats

You can create a format or informat using a Control data set. You can read about how to create formats using a control data set in the previous chapter. Here, you will see how to create an informat using this same technique.

If you want to create an informat rather than a format, the format name must start with an @ sign and the variable called Type must be set to "I" (stands for Informat). You can create a control data set from the Goals data set by renaming the variables and defining the Fmtname and Type variables. Here is the code:

Program 6.4: Using a control data set to create an informat

```
*Creating the INFORMAT using a CNTLIN data set;
data Control;
    set Goals(rename=(Year=Start Goal=Label));
    retain Fmtname '@goalfmt' Type 'I';
run;

proc format cntlin=Control;
    select @goalfmt;
run;
```

You want Year to be the start value (if there were a range instead of a single value, you would also have to define an End value), and Goal to be the label. An efficient way to assign the values of Fmtname and Type is with a RETAIN statement (as explained in the first task in Chapter 5).

Now that you have a properly structured control data set, run PROC FORMAT with the option CNTLIN=Control. The SELECT statement provides you with the following output:

```
       INFORMAT NAME: @GOALFMT LENGTH:   12    NUMBER OF VALUES:    9
    MIN LENGTH:    1   MAX LENGTH:   40   DEFAULT LENGTH:   12  FUZZ:        0
```

START	END	INVALUE(VER. V7|V8 12APR2012:10:47:22)
2004	2004	20
2005	2005	21
2006	2006	24
2007	2007	28
2008	2008	34
2009	2009	40
2010	2010	49
2011	2011	60
2012	2012	75

You can now use this INFORMAT to perform a table lookup as described in the previous task.

Task: Performing a one-way table lookup using a temporary array

Keywords

Temporary array

Table lookup

You can use a temporary array to store values for table lookup. This is one of my favorite methods. Temporary arrays are defined by an ARRAY statement with the keyword _TEMPORARY_ following the array name. Temporary arrays do not have any associated variables—they are simply subscripted locations, stored in memory (hence very fast) where you can store values. You can load a temporary array with data by assigning initial values when you define the array, or by loading the array from raw data or a SAS data set.

You can change the array bounds on temporary arrays, just as you do with regular arrays. In addition, you can have multi-dimensional temporary arrays so that you can perform table lookup based on two or more values. You will see an example of this at the end of this chapter.

For this example, you load a temporary array with the sales goals for the years 2004 to 2012 (the same as the previous examples). Once you do this, you can obtain the sales goal for any year by using the year as the array subscript. Here is the program:

Program 6.5: Using a temporary array to perform a table lookup

```
*Load a temporary with the Goals data;
data Sales_Goals;
   array Goalsarray[2004:2012] _temporary_;
   if _n_ = 1 then do Year = 2004 to 2012;
      set Goals;
      goalsarray[Year] = Goal;
   end;

   set Sales;
   Difference = Sales - Goalsarray[Year];
run;
```

The sales goals for the years 2004 to 2012 are stored in the Goals data set. You create a temporary array called Goalsarray and set the array bounds as 2004 to 2012. Following this, you enter the keyword _TEMPORARY_.

On the first iteration of the DATA step (when the automatic _N_ variable is equal to 1), you read in the nine sales goals into the nine array elements. The sales goal for any year is simply the array element with the subscript being the desired year.

Task: Performing a two-way table lookup using a temporary array

Keywords

Table lookup

Two-way table lookup

Temporary array

This example is similar to the previous one except now you have separate sales goals for each of four job categories. This information is stored in a SAS data set called Goals_Job and the first 10 observations are listed here:

Listing of the first 10 observations from data set Goals_Job

Year	Category	Goal
2004	1	20
2004	2	21
2004	3	24
2004	4	28
2005	1	34
2005	2	40
2005	3	49
2005	4	60
2006	1	75
2006	2	21

Once you load a two-dimensional temporary array with these values, you can obtain the sales goal for any year and job category by accessing the array element with the year as the first dimension and the job category as the second dimension. Here is the code:

Program 6.6: Demonstrating a two-way array to perform a table lookup

```
*Twoway table lookup using temporary arrays;
data Two_Way;
   array Goals_Job[2004:2012,4] _temporary_;
   if _n_ = 1 then do Year = 2004 to 2012;
      do Job = 1 to 4;
         set Goals_Job;
         Goals_Job[Year,Job] = Goal;
      end;
   end;

   set Sales_Job;
   drop Goal Job;
   Difference = Sales - Goals_Job[Year,Job];
run;
```

As in the previous program, you use a colon to define the bounds of the first dimension of the array (representing the year) and a comma to specify that there are four values for the second dimension (representing each of the four job codes). On the first iteration of the DATA step, you load values into the temporary array from the Goals_Job data set. Then, given any value for Year and Job, the array element returns the Goal value. The resulting data set Two_Way is listed next:

Year	Sales_ID	Sales	Difference
2004	1234	20.5	0.5
2005	1234	22.0	-12.0
2006	1234	26.0	-49.0
2007	1234	27.0	-3.0
2008	1234	37.0	-30.0
2009	1234	45.0	16.0
2010	1234	55.0	-7.0
2011	1234	61.0	23.0
2012	1234	72.0	12.0
2004	7477	18.0	-3.0

Using multi-dimensional temporary arrays is an easy and elegant way to perform table lookup.

Conclusion

Formats, informats, and temporary arrays are efficient ways to perform table lookup. Formats and informats seem to be one of the more popular methods used by SAS programmers. However, try using temporary arrays the next time you have a table lookup problem and you may become a fan of this technique.

Chapter 7: Restructuring (Transposing) SAS Data Sets

Introduction

At some point in your programming career you will need to restructure data sets. For example, you might have a data set that looks like this:

Listing of data set OnePer

Subj	Dx1	Dx2	Dx3
001	450	430	410
002	250	240	.
003	410	250	500
004	240	.	.

And you want to create a data set that looks like this:

Listing of data set ManyPer

Subj	Visit	Diagnosis
001	1	450
001	2	430
001	3	410
002	1	250
002	2	240
003	1	410
003	2	250
003	3	500
004	1	240

You might also want to reverse the process and go from a data set with several observations per subject to a data set containing one observation per subject. In this chapter, you will see how to accomplish these tasks using a DATA step and, if you are brave enough, PROC TRANSPOSE.

Task: Converting a data set with one observation per subject into one with multiple observations per subject (using a DATA step)

Keywords

Transpose

Restructure

Multiple observations per subject

This first task is to start with a data set that has one observation per subject, and create a data set that has multiple observations per subject. Let's start with the OnePer data set listed in the "Introduction" section. You can convert this data set into one with multiple observations per subject with the following program:

Program 7.1: Converting a data set with one observation per subject into one with multiple observations per subject (using a DATA step)

```
*One observation per subject to several observations per subject
 Data step approach;

data ManyPer;
   set OnePer;
   array Dx[3];
   do Visit = 1 to 3;
      if not missing(Dx[Visit]) then do;
         Diagnosis = Dx[Visit];
         output;
      end;
   end;
   keep Subj Diagnosis Visit;
run;
```

You first make an array of the three diagnosis variables (Dx1–Dx3). Next, you set up a DO loop to process each element in the array. If you have a missing Dx code, you do not want to output an observation in the ManyPer data set. You set the value of Diagnosis equal to Dx for non-missing values of Dx, and then output an observation to the ManyPer data set. It is important that the OUTPUT statement is inside the Do loop. In that way, you create multiple observations in data set ManyPer from a single observation in data set OnePer. You can see the listing of data set ManyPer in the "Introduction" section in this chapter.

Task: Converting a data set with one observation per subject into one with multiple observations per subject (using PROC TRANSPOSE)

Keywords

Transpose

Restructure

Multiple observations per subject

PROC TRANSPOSE

You can accomplish the same result as shown in the previous task by using PROC TRANSPOSE. If you run the following program, you will create a data set with multiple observations per subject; however, you will need to tweak it a bit to make the new data set look identical to the DATA step solution.

Program 7.2: Converting a data set with one observation into one with multiple observations per subject (using PROC TRANSPOSE)

```
*PROC TRANSPOSE solution;
*First try;
proc transpose data=oneper
               out=manyper;
   by Subj;
   var Dx1-Dx3;
run;
```

You supply PROC TRANSPOSE with an input data set (identified by DATA=) and an output data set (identified by OUT=). Because you want each of the rows of Dx codes to be transposed into a column of diagnosis codes for each Subj, you need to include a BY statement identifying Subj as your BY variable. You also identify the variables you want to transpose, in this case Dx1–Dx3. The resulting data set looks like this:

Subj	_NAME_	COL1
001	Dx1	450
001	Dx2	430
001	Dx3	410
002	Dx1	250
002	Dx2	240
002	Dx3	.
003	Dx1	410
003	Dx2	250
003	Dx3	500
004	Dx1	240
004	Dx2	.
004	Dx3	.

This is close to the result you want. You need to rename Col1 to Diagnosis, drop the variable _Name_, and remove observations that have a missing value for Dx, as demonstrated in the next program:

Program 7.3: Making a few changes to the PROC TRANSPOSE program to produce the output data set you want

```
proc transpose data=OnePer
                out=ManyPer(rename=(col1=Diagnosis)
                            drop=_name_
                            where=(Diagnosis is not null));
   by Subj;
   var Dx1-Dx3;
run;
```

You use a RENAME= data set option to rename Col1 to Diagnosis, a DROP = option to drop the variable _Name_, and a WHERE= data set option to remove observations where the Diagnosis is a missing value. The resulting data set is almost identical to the ManyPer data set produced by the DATA step—the only difference being that the variable Visit is not in the data set created by PROC TRANSPOSE.

Task: Converting a data set with multiple observations per subject into one with one observation per subject (using a DATA step)

Keywords

Transpose

Restructure

Multiple observations per subject

This task is the reverse of the first task described in this section. The approach, using a DATA step, is to create an array of the variables that will hold values from multiple observations in the input data set. Because you are adding one value to each array element for each iteration of the DATA step, you need to retain all of the variables making up the array. Be careful! If you do not set all of the new variables to a missing value for each new subject, you might wind up with non-missing values from a previous subject. The following program accomplishes this task:

Program 7.4: Converting a data set with multiple observations per subject into one with one observation per subject (using a DATA step)

```
*Going from a data set with several observations per subject
 to one with one observation per subject.;

proc sort data=ManyPer;
   by Subj Visit;
run;
```

```
data OnePer;
   set ManyPer;
   by Subj;
   array Dx[3];
   retain Dx1-Dx3;
   if first.Subj then call missing(of Dx1-Dx3);
   Dx[Visit] = Diagnosis;
   if last.Subj then output;
   keep Subj Dx1-Dx3;
run;
```

You need to sort the ManyPer data set by Subj in order to create the two automatic BY group variables First.Subj and Last.Subj. Each time you are reading the first observation for a subject, you set each of the Dx variables to a missing value (using the CALL MISSING routine). Next, you set each of the array elements equal to Diagnosis. Finally, when you are reading the last observation for each subject, you output an observation. The resulting data set is identical to the OnePer data set listed in the "Introduction" section in this chapter.

Task: Converting a data set with multiple observations per subject into one with one observation per subject (using PROC TRANSPOSE)

Keywords

Transpose

Restructure

Multiple observations per subject

The final task in this section is identical to the previous task, except you use PROC TRANSPOSE to perform the conversion. The program shown next has all of the options and data set options necessary to create the OnePer data set, which is identical to the previous task. Here it is:

Program 7.5: Converting a data set with multiple observations per subject into one with one observation per subject (using PROC TRANSPOSE)

```
proc transpose data=ManyPer out=OnePer(drop=_Name_)
   prefix=Dx;
   by Subj;
   id Visit;
   var Diagnosis;
run;
```

The PREFIX= option along with Visit as the ID variable tells PROC TRANSPOSE to combine the prefix (Dx) with the values of Visit (1,2,3) to create the three variables Dx1, Dx2, and Dx3. That is, the PREFIX= option and the ID statement interact to create the names of the transformed variables in the output data set. The VAR statement allows you to select which variables in your data set you want transposed. Since all the variables in the ManyPer data set have been identified in the BY and ID statements, you do not need to include the statement VAR DIAGNOSIS in this example. If you had other variables in the input data set, the VAR statement would have been necessary. It was included here since, in most real world problems, there would have been variables in the input data set that you did not want to transform.

Conclusion

Restructuring or transforming data sets, either from one-to-many or many-to-one is a task that most SAS programmers face at some point in their careers. Using a DATA step to accomplish your goals gives you more flexibility. If you use PROC TRANSPOSE frequently and don't have to rely on SAS help every time you use it (as does this author), you may find that you can solve your transposing problem faster with PROC TRANSPOSE compared to a DATA step. It's nice to have options.

Chapter 8: Tasks Involving Dates

Introduction

This short chapter includes one of the most common tasks—computing a person's age given a date of birth. Another task discussed is how to compute a SAS date given a month, day, and year. To make the task a bit more interesting, the program presented here uses the 15th day of the month to compute the date if month is a missing value. Let's get started.

Task: Computing a person's age, given his or her date of birth

Keywords

Age computation

YRDIF function

TODAY function

For this task, you want to compute a person's age as of January 1, 2012, given a date of birth (DOB). In the "old" days, you might approach this problem by taking the difference between January 1, 2012, and the DOB and dividing by 365.25. That almost works since there is a leap year every four years.

However, if you would like an exact solution, use the YRDIF function, introduced with SAS 9 and modified in SAS 9.3. Here is the program:

Program 8.1: Computing a person's age, given a date of birth

```
*Computing a person's age, given a date of birth;
data Compute_Age;
   set Demographic;
   Age_Exact = yrdif(DOB,'01jan2012'd);
   Age_Last_Birthday = int(Age_Exact);
run;
```

Data set Demographic contains the date of birth in a variable called DOB. The YRDIF function takes as its first two arguments, the earlier date and the later date, respectively. There is a third, optional argument called a basis that allows you to specify the number of days in a month and the number of days in a year. Some financial calculations require 30 day months and 360 day years, regardless of the actual months involved or if it is a leap year.

To use the YRDIF function prior to SAS 9.3, you need to specify "ACT/ACT" as the third argument to the function or you will get a syntax error. **Used in this manner, some of the age calculations may be off by one day if one or both of the dates involve a leap year**. Using SAS 9.3 or later, you can omit this third argument and the function defaults to a basis called "AGE" that computes ages correctly.

Once you have computed an age (or any other difference between two dates), you can use the INT function to compute the age as of the last birthday or the ROUND function to round the age to an integer (or other value).

If you want to compute an age as of the current date, substitute the TODAY function for the second argument of the YRDIF function.

Task: Computing a SAS date given a month, day, and year (even if the day value is missing)

Keywords

SAS Date

MDY function

COALESCE function

If you have a SAS data set with values of month, day, and year, you can use the MDY function to compute a SAS date. If the day of the month is missing, you might choose to use the 15th of the month for the day of the month.

The first program presented here uses a straightforward method of accomplishing this goal—the second program uses a clever and elegant method. Data set MoDayYear contains the variables Month, Day, and Year, but it does not contain a SAS date.

Here is the first program:

Program 8.2: Computing a SAS date given a month, day, and year value (even if the day of the month is missing)

```
*Creating a SAS date when the day of the month may be missing;
data Compute_Date;
   set MoDayYear;
   if missing(Day) then Date = MDY(Month,15,Year);
   else Date = MDY(Month,Day,Year);
   format Date date9.;
run;
```

I like this program because it uses one of my favorite functions, the MISSING function. This function can take character or numeric arguments. It returns a value of 1 (true) if the argument is missing, and a value of 0 (false) if the argument is not missing.

The next program uses the COALESCE function. You provide this function with a list of values and it returns the first value in the list that is not missing. I would like to thank Mark Jordan, an instructor for SAS, who thought up this solution, which is shown next:

Program 8.3: Computing a SAS date given a month, day, and year value (using the COALESCE function)

```
*Alternative (elegant) solution suggested by Mark Jordan;
data Compute_Date;
   set MoDayYear;
   Date = MDY(Month,coalesce(Day,15),Year);
   format Date date9.;
run;
```

The value of Day in this program is either the actual Day value or 15 if Day is a missing value.

Conclusion

Although this chapter only covered two tasks involving dates, you will find that these two tasks are essential to many of the program you write.

Chapter 9: Data Cleaning Tasks

Introduction

One of the first tasks facing a data analyst is to check for possible invalid numeric values. For some variables, you can decide if a value might be an error if it is outside a given range. For example, a value for resting heart rate would be unusual if it were below 40 or above 100.

Another approach for identifying possible numeric outliers is to see if a given data value does not seem to "belong" with the other values. A common approach is to compute a mean and standard deviation and declare a value as a possible outlier if it is a given number of standard deviations above or below the mean. Many statisticians use values of two or three standard deviations to identify possible outliers.

As you will see, automatic outlier detection, using a mean and standard deviation computed from the entire data set may not work if you have some large data errors. You will see how to overcome this problem so that you can use automatic outlier detection for many of your numeric variables.

Task: Looking for possible data errors using a given range

Keywords

Outliers

Range checking

If you can determine a reasonable range for some of your numeric variables, you can write a program like the following one to identify possible data errors:

Program 9.1: Looking for possible data errors using a given range

```
*Method using known ranges;
title "Listing of Patient Numbers and Invalid Data Values";
data _null_;
   set Blood_Pressure;
   file print;
   ***Check Heart_Rate;
   if (Heart_Rate lt 40 and not missing(Heart_Rate)) or
      Heart_Rate gt 100 then put Subj= @10 Heart_Rate=;
   ***Check SBP;
   if (SBP lt 80 and not missing(SBP)) or
      SBP gt 200 then put Subj= @10 SBP=;
   ***Check DBP;
   if (DBP lt 60 and not missing(DBP)) or
      DBP gt 120 then put Subj= @10 DBP=;
run;
```

This program reads data from the Blood_Pressure data set and prints out data for any subject whose heart rate, systolic blood pressure, or diastolic blood pressure is outside predetermined ranges. Output from this program follows:

```
Listing of Patient Numbers and Invalid Data Values
Subj=16   Heart_Rate=38
Subj=17   Heart_Rate=116
Subj=39   DBP=130
Subj=42   Heart_Rate=37
Subj=45   Heart_Rate=38
Subj=47   Heart_Rate=38
Subj=49   Heart_Rate=38
```

This solution is fine if you only have a few variables to test, but it can become tedious when you have a large number of variables.

Task: Demonstrating a macro to report on outliers using fixed ranges

Keywords

Outliers

Range checking

Errors macro

The macro, named Errors, presented here, allows you to specify lower and upper bounds for each numeric variable you want to test. In addition, you can determine if missing values should be flagged as errors. You run the macro once for each variable and, when finished, call another macro called Report (shown later) to print out a final report.

You can see an explanation of the macro following the listing:

Program 9.2: Presenting a macro for range checking of numeric variables

```
*Macro to perform range checking for numeric variables;
%macro Errors(Var=,      /* Variable to test     */
              Low=,      /* Low value            */
              High=,     /* High value           */
              Missing=IGNORE
                         /* How to treat missing values    */
                         /* Ignore is the default.  To flag */
                         /* missing values as errors set    */
                         /* Missing=error                   */);
data Tmp;
   set &Dsn(keep=&Idvar &Var);
   length Reason $ 10 Variable $ 32;
   Variable = "&Var";
   Value = &Var;
   if &Var lt &Low and not missing(&Var) then do;
      Reason='Low';
      output;
   end;
   %if %upcase(&Missing) ne IGNORE %then %do;
   else if missing(&Var) then do;
      Reason='Missing';
      output;
   end;
   %end;

   else if &Var gt &High then do;
      Reason='High';
      output;
      end;
      drop &Var;
   run;
```

```
   proc append base=Errors data=Tmp;
   run;
%mend errors;
```

To use the Errors macro, you first need to use a %LET statement to assign values to the two macro variables Dsn (the data set name) and Idvar (the name of the ID variable). Because you will be calling the macro several times, using %LET statements to assign these values saves you the trouble of reentering them each time you call the Errors macro.

You call the Errors macro with values for Var (the variable you want to test), Low (the lower bound), High (the higher bound), and Missing (with the default value of IGNORE) to ignore missing values, or with a value of ERROR (actually, any value not equal to IGNORE) to treat missing values as errors.

Each time you run this macro, observations containing errors are written out to the data set Tmp. Variables in this data set are Reason (High, Low, or Missing), Variable (the name of the variable being tested), and Value (the value of the variable being tested). You use PROC APPEND to add each of these observations to the data set Errors.

When you have run the Errors macro for all of the variables of interest, call the Report macro (listed in the following program) to generate a consolidated error report.

Program 9.3: Presenting the Report macro that generates a report after the Errors macro has been run

```
*Macro to generate an error report after the errors macro has been run;
%macro report;
   proc sort data=errors;
      by &Idvar;
   run;

   proc print data=errors;
      title "Error Report for Data Set &Dsn";
      id &Idvar;
      var Variable Value Reason;
   run;

   proc datasets library=work nolist;
      delete errors;
      delete tmp;
   run;
   quit;
%mend report;
```

This macro consists of a PROC SORT and a PROC PRINT. You run PROC DATASETS to delete the two temporary data sets (Tmp and Errors).

Suppose you want to perform the following tests on data set Blood_Pressure:

1. Heart_Rate is between 40 and 100 and missing values are to be considered errors.
2. SBP (systolic blood pressure) should be between 80 and 200 and missing values are to be ignored.
3. DBP (diastolic blood pressure) should be between 60 and 120 and missing values are to be ignored.

The ID variable in this data set is called Subj. Run the following lines of code to accomplish these goals:

Program 9.4: Calling the Errors and Report macros

```
*Test the macro;
%let Dsn = Blood_Pressure;
%let Idvar = Subj;
%errors(Var=Heart_Rate, Low=40, High=100, Missing=error)
%errors(Var=SBP, Low=80, High=200, Missing=ignore)
%errors(Var=DBP, Low=60, High=120, Missing=ignore)
%report
```

You call the Errors macro three times, once for each of your variables. You then call the Report macro to generate the final report, as shown next:

Subj	Variable	Value	Reason
2	Heart_Rate	.	Missing
8	Heart_Rate	.	Missing
16	Heart_Rate	38	Low
17	Heart_Rate	116	High
39	DBP	130	High
42	Heart_Rate	37	Low
45	Heart_Rate	38	Low
47	Heart_Rate	38	Low
49	Heart_Rate	38	Low

Notice that the two subjects with missing values for Heart_Rate are listed, but the missing values for SBP and DBP are ignored and are not listed in the report as errors.

Task: Demonstrating a macro that performs automatic outlier detection

Keywords

Outliers

Automatic outlier detection

Trimmed statistics

For some variables, it is possible to identify possible outliers, without specifying ranges. You can compute a mean and standard deviation and identify data points as possible outliers if they are more than a given number of standard deviations above or below the mean. There are several problems with this concept. First, if you have a highly skewed distribution, this technique may not work very well. Second, if you have a few extreme data errors (a misplaced decimal point, for example), the standard deviation might be so inflated that this technique will fail to identify data errors.

One method you can use to rectify this latter problem is to use trimmed statistics. Although this sounds complicated, the concept is really quite simple. That is, you throw away data values from both ends of the distribution before performing the calculations for the mean and standard deviation.

PROC UNIVARIATE has the ability to compute trimmed statistics. The macro that is presented next uses PROC UNIVARIATE to compute trimmed statistics, and outputs these values to a SAS data set using ODS. The output data set from the output object called TrimmedMeans contains the trimmed mean but it does not contain the trimmed standard deviation. Instead, you will find values for the trimmed standard error (the standard deviation divided by the square root of the sample size), and the degrees of freedom (in this case, equal to the sample size minus 1).

In addition to those slight problems, you need to restructure (transform) your original data set so that it can be merged with the ODS obtained data set.

The purpose of presenting the Auto_Outlier macro is to allow you to copy and use it. For those readers interested in the inner-workings of this macro, an explanation is included.

To call the Auto_Outlier macro, you need to identify your data set name, the ID variable, and a list of the variables you want to test (separated by spaces). Two additional, optional arguments are the amount of trimming you want and the number of standard deviations you want to use to declare data values as outliers. You have a choice on how to specify how much to trim your variables: If you enter a proportion (less than .5), this proportion will be trimmed from both ends of the distribution; if you specify an integer, this number of values will be trimmed from each end of the distribution.

A listing of the Auto_Outlier macro follows:

Program 9.5: Presenting a macro for automatic detection of possible outliers

```
*Method using automatic outlier detection;
%macro Auto_Outliers(
   Dsn=,        /* Data set name                        */
   ID=,         /* Name of ID variable                  */
   Var_list=,   /* List of variables to check           */
                /* separate names with spaces           */
   Trim=.1,     /* Integer 0 to n = number to trim      */
                /* from each tail; if between 0 and .5, */
                /* proportion to trim in each tail      */
   N_sd=2       /* Number of standard deviations        */);
   ods listing close;
   ods output TrimmedMeans=Trimmed(keep=VarName Mean Stdmean DF);
   proc univariate data=&Dsn trim=&Trim;
      var &Var_list;
   run;
   ods output close;

   data Restructure;
      set &Dsn;
      length Varname $ 32;
      array vars[*] &Var_list;
      do i = 1 to dim(vars);
         Varname = vname(vars[i]);
         Value = vars[i];
         output;
      end;
      keep &ID Varname Value;
   run;

   proc sort data=trimmed;
      by Varname;
   run;

   proc sort data=restructure;
      by Varname;
   run;

   data Outliers;
      merge Restructure Trimmed;
      by Varname;
      Std = StdMean*sqrt(DF + 1);
      if Value lt Mean - &N_sd*Std and not missing(Value)
         then do;
            Reason = 'Low  ';
            output;
         end;
      else if Value gt Mean + &N_sd*Std
         then do;
         Reason = 'High';
         output;
      end;
```

```
  run;

  proc sort data=Outliers;
     by &ID;
  run;

  ods listing;
  title "Outliers Based on Trimmed Statistics";
  proc print data=Outliers;
     id &ID;
     var Varname Value Reason;
  run;

  proc datasets nolist library=work;
     delete Trimmed;
     delete Restructure;
  run;
  quit;
%mend auto_outliers;
```

How the macro works

This section is included for those readers who want to understand the details of the Auto_Outlier macro. For this discussion, the macro was run on data set Blood_Pressure with the value of the macro variable &Var equal to "Heart_Rate SBP DBP".

The first step is to run PROC UNIVARIATE and capture the trimmed statistics in a SAS data set. You use an ODS OUTPUT statement to do this. Shown next is a listing of data set Trimmed:

VarName	Mean	StdMean	DF
Heart_Rate	69.3913	2.599517	45
SBP	132.0000	2.396861	43
DBP	79.7727	0.938919	43

You need to combine this information with the values from the Blood_Pressure data set. However, that data set has variables Subj, Heart_Rate, SBP, and DBP. So, your next step is to restructure the Blood_Pressure data set so you can merge it with the Trimmed data set. You accomplish this in DATA step Restructure. Refer to Chapter 7 for details on restructuring SAS data sets.

You may be unfamiliar with the VNAME function used in this DATA step. This very useful function returns a variable name, given an array element as the argument.

The first eight observations in data set Restructure (after it is sorted by Varname) are listed next:

Subj	Varname	Value
1	DBP	72
2	DBP	66
3	DBP	62
4	DBP	76
5	DBP	.
6	DBP	74
7	DBP	80
8	DBP	86

You can now merge the Trimmed data set and the Restructure data set using Varname as the BY variable. Data set Trimmed contains a standard error and the degrees of freedom. You compute the standard deviation by multiplying the standard error by the square root of the sample size (the degrees of freedom plus 1). Once you do this, you can compare each value with the mean plus or minus the specified number of standard deviations (all based on trimmed values).

Now, back to a discussion of how to use the Auto_Outlier macro. Let's test the macro using the three variables Heart_Rate, SBP, and DBP from the Blood_Pressure data set:

Program 9.6: Testing the Auto_Outliers macro

```
*Testing the auto outliers macro;
%auto_outliers(Dsn=Blood_Pressure,
               ID=Subj,
               Var_List=Heart_Rate SBP DBP,
               N_Sd=2.5)
```

This call is using the default value of .1 for the trim value, and a value of 2.5 (overriding the default value of 2.0) for the number of standard deviations. You can see the results of running this macro in the following listing:

Subj	Varname	Value	Reason
3	DBP	62	Low
12	SBP	188	High
16	SBP	196	High
17	Heart_Rate	116	High
24	SBP	180	High
30	SBP	82	Low
39	DBP	130	High
49	DBP	112	High
54	SBP	176	High

If you have highly skewed data or you suspect there are a lot of extreme data values, you might benefit from choosing a larger trim value. If you are running this macro on large data sets, you may also want to use larger values for the number of standard deviations.

Conclusion

Consider running one or both of the data cleaning macros presented in this chapter before you run any statistical tests on your data. You should also run PROC UNIVARIATE (including a HISTOGRAM statement) on all of the numeric variables.

Chapter 10: Reading Data with User-Defined INFORMATS

Introduction

Just about every SAS programmer can create and use formats. But not as many SAS programmers use PROC FORMAT to create their own informats. This is a very powerful feature of SAS. It is also quite easy to do—if you know how to write a format, you only need to learn a few things to be able to write an informat.

Task: Reading a combination of character and numeric data

Keywords

PROC FORMAT

Informat

Enhanced numeric informat

For this first task, you want to read peoples' oral temperatures. Instead of entering 98.6 (a normal body temperature), you want to enter an uppercase or lowercase 'n' (to save you some typing). The first program handles this problem using a traditional approach:

Program 10.1: Reading a combination of character and numeric data (traditional approach)

```
*A traditional approach to reading a combination
 of character and numeric data;
data Temperatures;
   input Dummy $ @@;
   if upcase(Dummy) = 'N' then Temp = 98.6;
   else Temp = input(Dummy,8.);
   if Temp gt 106 or Temp lt 96 then Temp = .;
   drop Dummy;
datalines;
101 N 97.3 111 n N 104.5 85
;
```

First, you read each data value as a character value. Next, you test if the value is an uppercase or lowercase 'n'. If so, you set the value of Temp to 98.6. If not, you use the INPUT function to perform the character-to-numeric conversion.

Here is a listing of data set Temperatures:

Temp
101.0
98.6
97.3
.
98.6
98.6
104.5
.

You can accomplish the same goal by creating your own informat, as follows:

Program 10.2: Reading a combination of character and numeric data (using a user-defined informat)

```
*Using an enhanced numeric informat to read
 a combination of character and numeric data;
proc format;
   invalue readtemp(upcase)
            96 - 106 = _same_
            'N'      = 98.6
            other    = .;
run;
```

```
data Temperatures;
    input Temp : readtemp5. @@;
datalines;
101 N 97.3 111 n N 67 104.5 85
;
```

You use a VALUE statement to make a format—you use an INVALUE statement to make an informat. Because the n's may be in lowercase, you use the UPCASE option (placed in parentheses following the informat name) to convert all letters to uppercase. Values between 96 and 106 are left alone (the keyword _SAME_ does this). A value of 98.6 (a numeric value) is used when the letter 'N' is read. All other values result in a missing value. An informat that reads both character and numeric data is referred to as an enhanced numeric informat. Data set Temperatures is identical to the one listed previously.

This is an efficient and elegant method for reading combinations of character and numeric data.

The next example reads a combination of numeric or letter grades and converts the letter grades to numeric values. Here is the program:

Program 10.3: Reading letter grades and converting them to numeric values using an informat

```
proc format;
    invalue readgrade(upcase just)
        'A' = 95
        'B' = 85
        'C' = 75
        'F' = 65
        other = _same_;
run;

data School;
    input ID : $3. Grade : readgrade3.;
datalines;
001 97
002 99
003  A
004 C
005 72
006   f
007 b
;
```

Two options, UPCASE and JUST, are used here. UPCASE, as described in the previous program, converts all letters to uppercase. The JUST option left justifies all text.

Here is a listing of data set School:

ID	Grade
001	97
002	99
003	95
004	75
005	72
006	65
007	85

As you can see, each student now has a numeric grade.

Conclusion

Writing your own informat is an easy and elegant approach to reading a combination of character and numeric data. Give it a try next time you are presented with this problem.

Chapter 11: Tasks Involving Multiple Observations per Subject

Introduction

This chapter demonstrates some of the techniques you need when you have multiple observations per subject (or any other grouping value). For example, you may have a SAS data set containing patient visits to a clinic. Each patient could have a different number of visits. Some of the tasks you may want to perform are selecting the first or last visit for each patient, computing differences in values from one visit to the next, or computing differences in values from the first visit to the last visit. Another common task is to determine the number of visits for each patient. You will find solutions to these tasks plus some other useful techniques in this chapter.

Task: Using PROC SORT to detect duplicate BY values or duplicate observations (records)

Keywords

PROC SORT

Duplicates

NODUPKEY

NODUPRECS

The data set Duplicates was created to demonstrate how to detect duplicate BY variables or duplicate observations in a SAS data set. Here is a listing:

Subj	Gender	Age	Height	Weight
001	M	23	63	122
002	F	44	59	109
002	F	44	59	109
003	M	87	67	200
004	F	100	53	112
004	F	50	59	201
005	M	45	69	188

Some subjects only have one observation. Subject 002 has a duplicate observation (the same value for all variables), while subject 004 has two observations with the same subject number but different values for the other variables.

You can use PROC SORT to detect both of these conditions and to remove either type of duplicate.

You can use the NODUPKEY option to remove any observations that have duplicate values on each of the BY variables. Suppose you want only one observation for each of the subjects in the Duplicates data set. Use the NODUPKEY option as follows:

Program 11.1: Demonstrating the NODUPKEY option of PROC SORT

```
*Using PROC SORT to detect duplicate BY values;
proc sort data=Duplicates out=Sorted nodupkey;
   by Subj;
run;
```

Since there is only one BY variable listed, the output data set (Sorted) will keep the first observation for each subject and remove all other observations with the same subject number. PROC SORT writes information about duplicate values in the Log, as shown here:

```
NOTE: There were 7 observations read from the data set WORK.DUPLICATES.
NOTE: 2 observations with duplicate key values were deleted.
NOTE: The data set WORK.SORTED has 5 observations and 5 variables.
```

You see here that two observations with duplicate key (BY) values were deleted. Here is the listing:

Subj	Gender	Age	Height	Weight
001	M	23	63	122
002	F	44	59	109
003	M	87	67	200
004	F	100	53	112
005	M	45	69	188

The next program demonstrates the NODUPRECS option.

The NODUPRECS option (also called NODUP) removes duplicate observations, that is, observations that have all the same value for all of the variables. In previous versions of SAS, this option was called NODUP and you can still use that term. However, this author believes that it is better to use NODUPRECS since it makes it clearer to someone reading the program that you are removing duplicate observations (records).

The next program demonstrates this option:

Program 11.2: Demonstrating the NODUPRECS option of PROC SORT

```
*Using PROC SORT to detect duplicate records;
proc sort data=Duplicates out=Sorted noduprecs;
   by Subj;
run;
```

The resulting SAS Log and output are shown here:

```
NOTE: There were 7 observations read from the data set WORK.DUPLICATES.
NOTE: 1 duplicate observations were deleted.
NOTE: The data set WORK.SORTED has 6 observations and 5 variables.
```

This time only one observation (from Subj 002) was removed. Here is the listing:

Subj	Gender	Age	Height	Weight
001	M	23	63	122
002	F	44	59	109
003	M	87	67	200
004	F	100	53	112
004	F	50	59	201
005	M	45	69	188

There is a possible problem when you use the NODUPRECS option. First take a look at a listing of data set Multiple:

Subj	X	Y
001	**1**	**2**
001	3	2
001	**1**	**2**
002	5	7
003	**7**	**8**
003	**7**	**8**
005	4	5

All of the observations in **bold** represent duplicate observations. Let's see what happens when you run PROC SORT with the NODUPRECS option:

Program 11.3: Demonstrating a possible problem with the NODUPRECS option

```
*Demonstrating a feature of noduprecs;
proc sort data=Multiple out=Features noduprecs;
   by Subj;
run;
```

Here is the surprising result:

Listing of data set Features

Subj	X	Y
001	**1**	**2**
001	3	2
001	**1**	**2**
002	5	7
003	7	8
005	4	5

The duplicate observation for subject 003 was removed, but there is still a duplicate observation for subject 001. What's going on?

If you read the documentation on the NODUPRECS option of PROC SORT, you find that the NODUPRECS option removes **successive** duplicates. What exactly does this mean?

When you sorted the observations by subject, the two duplicate observations for subject 001 were not successive. An easy solution to ensure that all duplicate observations are successive is to sort by all of the BY variables. So, if you rewrite Program 11.3 like this:

Program 11.4: Fixing the problem with the NODUPRECS option

```
*Possible solution to the problem;
proc sort data=Multiple out=Features noduprecs;
   by _all_;
run;
```

You will get the desire result as follows:

Subj	X	Y
001	1	2
001	3	2
002	5	7
003	7	8
005	4	5

In practice, when you have a large data set with a large number of variables, you may not want to sort by _ALL_. If you simply list several BY variables, especially ones that have lots of different values, you can be confident that any duplicate observations are successive.

Task: Extracting the first and last observation in a BY group

Keywords

First. variables

Last. variables

SAS programs are observation oriented. That is, you process one observation at a time. When you are done, a new observation enters the PDV (program data vector) and the previous observation is gone (unless some variables' values were retained, either by default or due to a RETAIN statement). To perform operations involving multiple observations per subject, SAS provides you with some useful tools. The two logical variables First.*by-variable-name* and Last.*by-variable-name* are created when you follow a SET statement with a BY statement.

These are temporary variables that exist only while the DATA step is processing, and they allow you to determine when you are reading the first observation in a BY group or the last observation in a BY group. The next program demonstrates how this works.

Program 11.5: Demonstrating First. and Last. variables

```
proc sort data=Duplicates out=Sorted_Duplicates;
   by Subj;
run;

data _null_;
   set Sorted_Duplicates;
   by Subj;
   put Subj= First.Subj= Last.Subj=;
run;
```

You first sort your data set by one or more BY variables. In this example, you sort the Duplicates data set by Subj. In the data _NULL_ DATA step, you follow the SET statement with a BY statement. This creates the two logical variables First.Subj and Last.Subj. As you might expect, First.Subj is true when you are reading the first observation for a subject, and Last.Subj is true when you are reading the last observation for a subject. A value of 1 indicates true; a value of 0 indicates false.

Here is the portion of the SAS Log that shows the values of these two variables:

```
Subj=001 FIRST.Subj=1 LAST.Subj=1
Subj=002 FIRST.Subj=1 LAST.Subj=0
Subj=002 FIRST.Subj=0 LAST.Subj=1
Subj=003 FIRST.Subj=1 LAST.Subj=1
Subj=004 FIRST.Subj=1 LAST.Subj=0
Subj=004 FIRST.Subj=0 LAST.Subj=1
Subj=005 FIRST.Subj=1 LAST.Subj=1
```

First. and Last. variables are extremely useful. You may need to perform some operations when you are reading the first observation for a subject (such as initializing counters), and also perform certain operations when you are reading the last observation for a subject (such as outputting an observation).

As an example, if you want to output the last observation for each subject, proceed as follows:

Program 11.6: Selecting the last observation for each subject

```
proc sort data=Duplicates out=Sorted_Duplicates;
    by Subj;
run;

data Last;
    set Sorted_Duplicates;
    by Subj;
    if Last.Subj;
run;
```

The statement BY Subj creates the two variables First.Subj and Last.Subj. The subsetting IF statement is only true when you are reading the last observation for each subject (it is also equivalent to: if Last.Subj = 1), and the resulting data set now contains the last observation for each subject as shown in the following listing:

Subj	Gender	Age	Height	Weight
001	M	23	63	122
002	F	44	59	109
003	M	87	67	200
004	F	50	59	201
005	M	45	69	188

You can also use a similar technique to output the first observation for each subject.

Task: Detecting duplicate BY values using a DATA step

Keywords

PROC SORT

Duplicate BY values

First. variables

Last. variables

You can use the First. and Last. variables described in the previous section to perform tasks involving duplicates and multiple observations per subject.

If you want to determine if there are any duplicate values for Subj in the Duplicates data set (and output them to a SAS data set), proceed as follows:

Program 11.7: Detecting duplicate BY values using a DATA step

```
*Using a DATA step to detect duplicate BY values;
proc sort data=Duplicates out=Sorted_Duplicates;
   by Subj;
run;

data Dups;
   set Sorted_Duplicates;
   by Subj;
   if first.Subj and last.Subj then delete;
run;
```

If there is only one observation for a value of Subj, then both First.Subj and Last.Subj will be true. If you delete these observations, you are left with a data set of duplicate BY values, as shown in the following listing:

Subj	Gender	Age	Height	Weight
002	F	44	59	109
002	F	44	59	109
004	F	100	53	112
004	F	50	59	201

Task: Identifying observations with exactly 'n' observations per subject

Keywords

Multiple observations per subject

"n" observations per subject

There are times when you expect a fixed number of observations per subject (or other BY variable). You can accomplish this task by initializing a counter when you are reading the first observation per subject, incrementing the counter for each successive observation with the same subject number, and checking the counter when you are reading the last observation for each subject.

Data set TwoRecords was created to demonstrate this task. Here is the listing:

Subj	Weight
001	200
001	190
002	155
002	157
003	123
004	220
004	221
004	210
005	111
005	112

Subjects 001, 002, and 005 have two observations each. You can output the observations for all subjects who do not have exactly two observations with the following program:

Program 11.8: Identifying subjects who have exactly two observations

```
proc sort data=Two_Records;
   by Subj;
run;

data Not_Two;
   set Two_Records;
   by subj;
   if first.Subj then n=0;
   n + 1;
   if last.Subj and n ne 2 then output;
run;
```

Each time you are processing the first observation for a subject, you set your counter (n) equal to zero. You then increment the counter using a SUM statement. Finally, when you reach the last observation for each subject and n is not equal to 2, you output an observation.

The following listing shows all the subjects who did not have exactly two observations:

Subj	Weight	n
003	123	1
004	210	3

Task: Computing differences between observations (for each subject)

Keywords

Differences between observations

LAG function

One of the most common requirements when you have multiple observations per subject is to compute differences between observations. To demonstrate this, take a look at the Visits data set. (Each subject has a unique value of the variable Patient.)

Patient	Visit	Weight
001	1	120
001	2	124
001	3	124
002	1	200
003	1	310

Patient	Visit	Weight
003	2	305
003	3	298
003	3	290
004	1	160
004	2	162

Each patient has from one to three visits. You want to see the weight change from visit to visit for each patient. The simplest way is to use the LAG function, like this:

Program 11.9: Using the LAG function to compute differences between observations

```
*Computing inter-patient differences;
proc sort data=Visits;
   by Patient Visit;
run;

data Difference;
   set Visits;
   by Patient;
   Diff_Wt = Weight - lag(Weight);
   if not first.Patient then output;
run;
```

The LAG function returns the value of its argument the last time the function executed. **If you execute the LAG function for every iteration of the DATA step, it will return a value from the previous observation.** You may be tempted not to execute the LAG function when you are reading the first observation for each patient, since the lag value is the last value from the previous patient. This is OK. You are only outputting differences for the second through last observation for each patient. If you use a statement such as:

```
if not First.Patient then Diff_Wt = Weight - lag(Weight);
```

it will not work because the next time you execute the LAG function it will still return the last value from the previous patient. **Unless you are using the LAG function in a very unusual situation, such as in Program 11.11, you should not use the LAG function in conditional logic.**

In the previous program, you can use the DIF function (DIF(Weight)) instead of the LAG function (equal to Weight – LAG(Weight) if you prefer.

Here is a list of data set Difference:

Patient	Visit	Weight	Diff_Wt
001	2	124	4
001	3	124	0
003	2	305	-5
003	3	298	-7
003	3	290	-8
004	2	162	2

Task: Computing the difference between the first and last observation for each subject

Keywords

Differences between first and last observations

RETAIN statement

LAG function (executed conditionally)

A similar task to the previous task is to compute differences between the first and last visit for each patient. There are two interesting ways to accomplish this: The first method uses retained variables (ones that are not set back to missing for each iteration of the DATA step). The other method uses the LAG function in a very unconventional way.

Here is the first program, followed by an explanation:

Program 11.10: Computing the differences between the first visit and last visit for each patient

```
*Computing the differences between the first and last visit;
proc sort data=Visits;
   by Patient Visit;
run;

data First_Last;
   set Visits;
   by Patient;
   *Delete observations where only one visit;
   if first.Patient and last.Patient then delete;
   retain First_Wt;
   if first.Patient then First_Wt = Weight;
```

```
   if last.Patient then do;
      Diff_Wt = Weight - First_Wt;
      output;
   end;
run;
```

You only need the data set sorted by Patient to create the First.Patient and Last.Patient variables. However, it is a good idea, while you are sorting, to include Visit in the sort to ensure that all the visits for each patient are also in order (don't assume things when you don't have to).

Next, you remove any patient with only one observation—you can't compute differences on these patients. You accomplish this by deleting all observations where First.Patient and Last.Patient are both true. Next, you use a RETAIN statement for the variable First_Wt. When you read the first observation for any patient, you set First_wt equal to Weight. When you are reading the last visit for any patient, you compute the difference of First_Wt and Weight.

A listing of data set First_Last is shown here:

Patient	Visit	Weight	First_Wt	Diff_Wt
001	3	124	120	4
003	3	290	310	-20
004	2	162	160	2

An interesting and unconventional program to solve this same task is shown next:

Program 11.11: Using the LAG function to compute differences between the first and last visit for each patient

```
*Redoing the previous program using the LAG function;
data _First_Last;
   set Visits;
   by Patient;
   *Delete observations where only one visit;
   if first.Patient and last.Patient then delete;
   if first.Patient or last.Patient then Diff_Wt = Weight -
lag(Weight);
   if last.Patient then output;
run;
```

Remember, earlier in this chapter you were cautioned not to execute the LAG (or DIF) function conditionally (that is, following an IF statement). However, for this application it works. You first eliminate any patient with only one visit (as in the previous program). Next, you execute the LAG function only when you are reading the first **or** last observation for each patient. When you are reading a value from the last observation for a patient, you need to realize that the last time the LAG function executed was at the first visit. Therefore, you are obtaining a difference between the first and last visits.

Conclusion

Even though SAS is an observation-based language, it has all the tools you need when analyzing data sets with multiple observations per subject.

Chapter 12: Miscellaneous Tasks

Introduction

Because the word miscellaneous is in the title of this chapter, don't be misled into thinking these tasks are not useful—it's just that they do not fit in any convenient category.

Task: Determining the number of observations in a SAS data set (using the NOBS= SET option)

Keywords

Number of observations

NOBS

There are several ways to determine the number of observations in a SAS data set. One of the most common methods is to use the NOBS=*variable_name* option on a SET statement. At compile time, *variable_name* is set equal to the number of observations in the data set named on the SET statement.

In this example, you want to determine the number of observations in the Stocks data set.

Program 12.1: Using the NOBS= SET option to determine the number of observations in a SAS data set

```
*Determining the number of observations in a SAS data set;
*Using the SET option NOBS=;

data New;
   set Stocks nobs=Number_of_obs;
   if _n_ =1 then
      put "The number of observations in data set STOCKS is: "
          Number_of_obs;
   How_Far = _n_ / Number_of_obs;
run;
```

You use the NOBS= option on the SET statement to obtain the number of observations in the Stocks data set (and assign this value to Number_of_Obs). In this example, you want to compute the observation number divided by the total number of observations in the Stocks data set. You also use a PUT statement to output this number to the SAS Log. A listing of the first eight observations in data set New is displayed here:

Date	Price	How_Far
02JAN2012	24	0.02381
03JAN2012	24	0.04762
04JAN2012	33	0.07143
05JAN2012	25	0.09524
06JAN2012	25	0.11905
09JAN2012	24	0.14286
10JAN2012	33	0.16667
11JAN2012	34	0.19048

A section from the SAS Log shows that there were 42 observations in the Stocks data set:

```
The number of observations in data set STOCKS is: 42
```

Task: Determining the number of observations in a SAS data set and assigning the value to a macro variable

Keywords

Number of observations

NOBS

You may want to set a macro variable equal to the number of observations in a SAS data set. Take a look at the rather interesting program shown next:

Program 12.2: Placing the number of observations in a SAS data set into a macro variable

```
*Putting the number of observations in a SAS data set into
 a macro variable;
data _null_;
   if 0 then set Stocks nobs=Number_of_obs;
   call symputx('N_of_obs',Number_of_obs);
   put "The number of observations in STOCK is &N_of_obs";
   stop;
run;
```

Yes, the IF statement definitely looks strange! Zero is always false, so the SET statement never executes. However, the NOBS= option does its magic at compile time. CALL SYMPUTX assigns the value of Number_of_obs to the macro variable N_of_obs. Since this DATA step is not reading any observations from the Stocks data set, you need to include the STOP statement. (Normally, a DATA step stops when you reach the end-of-file on any data set.)

You can now use the macro variable in other DATA steps. For example, to list the last five observations in the Stocks data set, you could use the following program:

Program 12.3: Listing the last five observations in data set Stocks

```
%let Start = %eval(&N_of_obs - 4);

data Last_Five;
   set Stocks(firstobs=&Start);
run;
```

The %LET statement creates another macro variable (Start) that is equal to the number of observations in the Stocks data set minus 4. You need to use the %EVAL function to compute the Start value because all macro values are text and you cannot subtract 4 from a text value (42). The %EVAL function tells the program to treat each macro value as a number and perform the specified arithmetic.

Using this value of Start with the FIRSTOBS= data set option gives you the last five observations in the data set. Here is the listing:

Date	Price
22FEB2012	31
23FEB2012	31
24FEB2012	33
27FEB2012	37
28FEB2012	33

Task: Determining the number of observations in a SAS data set (using library tables)

Keywords

Number of observations

Library tables

SAS library tables contain metadata relating to your data sets, and you can obtain this information from SASHELP.VTABLE by specifying the LIBNAME and the name of the data set you are interested in. This method of determining the number of observations in a SAS data set has an advantage over the previous methods described so far. That is, if you (or other people) are modifying a data set, you need to know the total number of observations in a data set as well as the number of observations that have been marked for deletion (but are still counted when you use the NOBS= SET option).

The next program demonstrates how to use library tables to determine the number of observations in the Stocks data set:

Program 12.4: Determining the number of observations in a data set using library tables

```
*Method using library tables;
data _null_ ;
  set sashelp.vtable;
  where libname="WORK" and memname="STOCKS" ;
  ***Note: you must use uppercase values for libname and memname;
  Num_obs=Nobs-Delobs;
  ***Nobs is total number of observations, including deleted ones
     Delobs is the number of deleted observations;
  put "Nobs - Delobs: num_obs = " num_obs;
run;
```

It is important to remember that the libname and memname must be entered in uppercase, because that is the way they are stored in the table. The two variables you need from the table are Nobs (the total number of observations including ones marked for deletion) and Delobs (the number of observations marked for deletion). When you run this program, the line produced by the PUT statement is:

```
nobs - delobs: num_obs = 42
```

Task: Determining the number of observations in a SAS data set (using SAS functions)

Keywords

Number of observations

Exist function

Attrn function

What, another way to determine the number of observations in a SAS data set? A group of functions in the category File I/O allow you to obtain information about a SAS data set.

Here is the program, followed by an explanation:

Program 12.5: Determining the number of observations in a SAS data set using SAS functions

```
*Using SAS functions;
data _null_;
   Exist = exist("stocks");
   if Exist then do;
      Dsid = open("Stocks");
      Num_obs = attrn(Dsid,"Nlobs");
      *Nlobs is the number of logical observations,
       observations not marked for deleting;
      put "Number of observations in STOCK is: " Num_obs;
   end;
   else put "Data set STOCKS does not exist";
   RC = close(Dsid);
run;
```

The EXIST function returns a value of true if the data set exists, and a value of false if it does not exist. Next, you open the data set using the OPEN function. This function returns a data set ID that you can use with other file I/O functions. You enter the data set ID as the first argument of the ATTRN function, and any one of several character values for the second argument of this function. NLOBS gives you the number of logical observations (observations not marked for deletion). Another useful value for this second argument is NVAR, which returns the number of variables in your data set.

The line in the SAS Log produced by the PUT statement in this program is:

```
Number of observations in STOCK is: 42
```

It is recommended that you close the data set (using the CLOSE function) after you are finished.

Task: Counting the number of a specific response in a list of variables

Keywords

CATS function

COUNTN function

Counting specific values in a list of variables

For this example, you have a data set (Questionnaire) with five variables, Q1–Q5. Each variable is a 'Y' or 'N' (yes or no) value and these values may be in uppercase or lowercase. The task is to count the number of Y's and N's. This first program demonstrates the "old fashioned" way to accomplish this goal. Here is the program:

Program 12.6: Counting the number of Y's and N's in a list of SAS variables (using arrays and loops)

```
*Counting the number of a specific response in a list of variables;
*"Old Fashioned" solution;

data Count_YN;
   set Questionnaire;
   array Q[5];
   Num_Y = 0;
   Num_N = 0;
   do i = 1 to 5;
      if upcase(Q[i]) eq 'Y' then Num_Y + 1;
      else if upcase(Q[i]) eq 'N' then Num_N + 1;
   end;
   drop i;
run;
```

You first create an array of the five variables. Next, you initialize the two variables that will hold the counts, Num_Y and Num_X, to zero. Finally, you use a DO loop to check the value of each variable and check if it is a 'Y' or an 'N'. The UPCASE function is needed since some of the values may be in lowercase.

A listing of the Count_YN data set is shown next:

Subj	Q1	Q2	Q3	Q4	Q5	Num_Y	Num_N
1	y	y	n	n	y	3	2
2	N	N	n	Y	3	1	3
4	y	Y	n	N	y	3	2
5	y	y	y	y	y	5	0

A much more elegant solution to this task is to use two SAS functions, CATS and COUNTC. The CATS function takes each of the arguments, strips leading and trailing blanks, and concatenates the strings. The COUNTC function counts the number of times a particular character (or characters) appears in a string. Together, these functions make for a very compact and elegant solution, as shown here:

Program 12.7: Using the CATS and COUNTC functions to count responses

```
*Solution using the Count and Cats functions;
data Count_YN;
   set Questionnaire;
   Num_Y = countc(cats(of Q1-Q5),'Y','i');
   Num_N = countc(cats(of Q1-Q5),'N','i');
run;
```

The first argument of the COUNTC function is the string you want to query, the second argument specifies one or more characters that you want to count, and the third argument is a modifier. The 'I' modifier means, ignore case. Running this program produces the same output as the longer program described previously.

Task: Computing a moving average

Keywords

Moving average

LAG function

A moving average is used to see trends in data that might have large fluctuations. In this example, you want to average three days of stock prices to see a trend that may be hard to see because of the large day-to-day fluctuations. The first average is the average of days 1, 2 and 3, the second average is the average of days 2, 3, and 4 and so forth. Since each daily price is in a separate observation (sorted by date), you can use the LAG function to obtain values from previous observations.

Program 12.8: Computing a moving average

```
*How to compute a moving average;
data Moving;
   set Stocks;
   Last = lag(Price);
   Twoback = lag2(Price);
   if _n_ ge 3 then Moving = mean(of Price,Last,Twoback);
run;
title "Plots of Stock Price and Three Day Moving Average";
proc sgplot data=moving;
   series x=Date y=Price;
   series x=Date y=Moving;
run;
```

Since you execute the LAG (and LAG2) functions for every iteration of the DATA step, the LAG value is the price from the previous observation, and the LAG2 value is the price from the observation before that. You could start outputting averages from the start, but in this example, you start outputting observations only when you get to the third observation (when _n_ is greater than or equal to 3).

To help see the difference between the individual data points and the moving average, you can use PROC SGPLOT to see both plots together, as shown next:

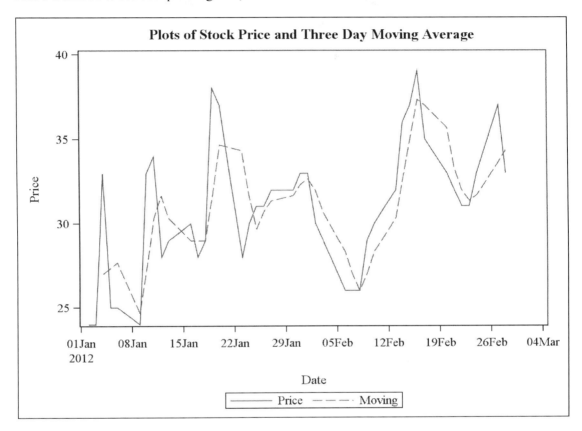

As you can see, the dashed line, representing the moving average, does not fluctuate as much as the raw data values.

Task: Presenting a macro to compute a moving average

Keywords

Moving average

Moving average macro

If you frequently need to compute moving averages or if you want the moving average of a large number of data points, you may wish to use the macro presented here.

You call the macro (Moving_Ave) with the name of your input data set, the name of the new data set that will include all the variables from the input data set along with the moving average, the name of the variable to be averaged, and how many observations to include in each average.

Following the macro is a brief explanation of how it works:

Program 12.9: Presenting a macro to compute a moving average

```
*Presenting a macro to compute a moving average;

%macro Moving_ave(In_dsn=,       /*Input data set name          */
                  Out_dsn=,      /*Output data set name         */
                  Var=,          /*Variable on which to compute
                                    the average                 */
                  Moving=,       /* Variable for moving average */
                  n=             /* Number of observations on which
                                    to compute the average       */);
   data &Out_dsn;
      set &In_dsn;
        ***compute the lags;
      _x1 = &Var;
      %do i = 1 %to &n - 1;
         %let Num = %eval(&i + 1);
           _x&Num = lag&i(&Var);
      %end;

        ***if the observation number is greater than or equal to the
           number of values needed for the moving average, output;
   if _n_ ge &n then do;
      &Moving = mean (of _x1 - _x&n);
      output;
   end;
   drop _x: ;
   run;
%mend Moving_ave;
```

You start with a DATA step that brings in each observation from the input data set (&In_dsn). Next, you need to assign names for the original variable and each of its 'n' lag values. You use the variable name _x1 for the non-lagged value of your variable, _x2 for the LAG1 value, _x3 for the LAG2 value and so forth. The value of _x1 is set outside the %DO loop and each of the other _x values are assigned by concatenating the prefix _x with the value of the %DO loop counter plus 1. For example, in the first iteration of the %DO loop, i is equal to 1, Num is equal to 2, and the assignment statement resolves as _x2 = lag1(*variable*). Once you have computed all of the lag values, you use the MEAN function to compute the moving average.

To get a better feel for how this works, let's call the macro like this:

Program 12.10: Calling the Moving_ave macro

```
*Testing the macro;
%moving_Ave(In_dsn=Stocks,
            Out_dsn=Moving_stocks,
            Var=Price,
            Moving=Average,
            n=5)
```

and then inspect the SAS log to see the macro generated code (accomplished by turning on the MPRINT option). Here is the generated code:

```
MPRINT(MOVING_AVE):   data Moving_stocks;
MPRINT(MOVING_AVE):   set Stocks;
MPRINT(MOVING_AVE):   ***compute the lags;
MPRINT(MOVING_AVE):   _x1 = Price;
MPRINT(MOVING_AVE):   _x2 = lag1(Price);
MPRINT(MOVING_AVE):   _x3 = lag2(Price);
MPRINT(MOVING_AVE):   _x4 = lag3(Price);
MPRINT(MOVING_AVE):   _x5 = lag4(Price);
MPRINT(MOVING_AVE):   ***if the observation number is greater than or
equal to the number of values needed for the moving average, output;
MPRINT(MOVING_AVE):   if _n_ ge 5 then do;
MPRINT(MOVING_AVE):   Average = mean (of _x1 - _x5);
MPRINT(MOVING_AVE):   output;
MPRINT(MOVING_AVE):   end;
MPRINT(MOVING_AVE):   drop _x: ;
MPRINT(MOVING_AVE):   run;
```

Remember that you can copy this macro and all the programs in this book by going to the SAS web site (support.sas.com/cody).

Task: Replacing the first eight digits of a credit card number with asterisks

Keywords

Masking characters

SUBSTR function

CATS function

SUBSTR function on the left-hand side of the equal sign

This task is not limited to masking digits in credit card numbers—you can use this technique any time you want to replace characters in a string. You will see two solutions to this task: One simple and straightforward, the other, very elegant.

In this simple solution, you use the SUBSTR function to extract the last 4 digits of the credit card number and you follow this with the CATS function to concatenate eight asterisks with these last four digits. Here is the code:

Program 12.11: Replacing the first eight digits of a credit card number with asterisks (method 1)

```
*Replacing the first 8 digits of a credit card number
 with asterisks;

*Using substrings and concatenation functions;
data Credit_Report;
   length Last_Four $ 12;
   set Credit;
   Last_Four = cats('********',substr(Account,9));
run;
```

A more elegant solution is to use the SUBSTR function on the left-hand side of the equal sign like this:

Program 12.12: Replacing the first eight digits of a credit card number with asterisks (method 2)

```
*Using the SUBSTR function on the left-hand side of the equal sign;
data Credit_report;
   set Credit;
   Last_Four = Account;
   substr(Last_Four,1,8) = '********';
run;
```

As a review, the SUBSTR function on the left-hand side of the equal sign places the characters to the right of the equal sign into the existing variable named on the left-hand side of the equal sign, starting at the position specified by the second argument of the function for a length specified by the third argument of the function. Yes, this program is only one line shorter than the previous program, but this author considers it a more elegant solution—and isn't that your goal—to write elegant programs?

Task: Sorting within an observation (using the ORDINAL function)

Keywords

Sorting within an observation

Ordinal function

There are numerous occasions where you would like to sort values within an observation. For example, you might have a number of scores per person and you want to order the scores from lowest to highest. The first solution presented here, uses the ORDINAL function to accomplish this goal. This function takes a number as the first argument, and a list of values as the remaining arguments. When the number is equal to one, the function returns the smallest value, when the number is equal to two, the function returns the second smallest value, and so on. Note that this function does not ignore missing values, so the smallest value may be a missing value. Another ordering function, SMALLEST, does ignore missing values—it returns the smallest nth value in a list of values, not counting any missing values.

You want to order 10 scores (Score1–Score10) from lowest to highest. These scores are stored in data set Scores, as shown below:

Name	Score1	Score2	Score3	Score4	Score5	Score6	Score7	Score8	Score9	Score10
John	95	92	87	100	96	88	89	78	2	95
Mary	98	96	93	89	95	95	94	.	.	99
Sarpal	87	84	87	88	80	.	81	78	77	92
Sophie	78	79	81	82	84	85	86	88	90	95

For this task, you want to create 10 new variables (Sorted1–Sorted10) equal to the 10 scores sorted from smallest to largest.

The following program accomplishes this task:

Program 12.13: Using the ORDINAL function to sort within an observation

```
*Sorting within an observation: Using the ORDINAL function;
data Ordered_Scores;
   set Scores;
   array Score[10];
   array Sorted[10];
   do i = 1 to 10;
      Sorted[i] = ordinal(i, of Score[*]);
   end;
   drop i;
run;
```

You may wonder why there are no variable names listed on the two ARRAY statements. If you define an array with the number of elements in the brackets following the array name, and you do not include a list of variables, SAS will use the array name with numbers from 1 to n appended. For example, array Score has 10 variables, Score1–Score10. This program also uses Score[*] as the second argument to the ORDINAL function. This is equivalent to writing Score1–Score10.

As you iterate the DO loop, the values of Score are placed in the Sorted variables from smallest to largest.

Task: Sorting within an observation (using CALL SORTN)

Keywords

Sorting within an observation

CALL SORTN

CALL SORTN is a call routine that was introduced in SAS 9.3. (Note: CALL SORTN was actually in earlier versions of SAS 9, but considered experimental.)

For this example, you want to order the variables Score1–Score10 from smallest to largest. To use the SORTN routine, you enter a list of variable names as arguments to the routine. If you have a list in the form Score1–Score10, you need to precede the list with the keyword OF. Following the call, the values of all the variables listed in the routine are ordered from smallest to largest.

Here is an example:

Program 12.14: Sorting within an observation (using CALL SORTN)

```
*Sorting within an observation: Using CALL SORTN;
data Ordered_Scores;
   set Scores;
   call sortn(of Score1-Score10);
run;
```

After this program executes, the values for Score1–Score10 have been placed in order from smallest (Score1) to largest (Score10) as shown in the following listing:

Name	Score1	Score2	Score3	Score4	Score5	Score6	Score7	Score8	Score9	Score10
John	2	78	87	88	89	92	95	95	96	100
Mary	.	.	89	93	94	95	95	96	98	99
Sarpal	.	77	78	80	81	84	87	87	88	92
Sophie	78	79	81	82	84	85	86	88	90	95

If you also want to keep the original values of Score1–Score10 in their original order, you need to copy these values to other variables before calling SORTN. It might be easier to use the ORDINAL function solution if you need the original order and the sorted order values.

Task: Computing the average of the 'n' highest scores

Keywords

Sorting within an observation

CALL SORTN

As a practical example of why it is useful to sort values within an observation, consider the following: You have 10 scores (Score1–Score10) and you want to compute the average of the eight highest scores.

One solution that comes to mind, is to use the ORDINAL function to set the two lowest values to missing, and then use the MEAN function to compute the average.

CALL SORTN provides a very simple way to solve this problem.

Program 12.15: Computing an average of the 'n' highest scores (using CALL SORTN)

```
*Sorting within an observation: Using CALL SORTN;
data Ordered_Scores;
   set Scores;
   call sortn(of Score1-Score10);
   Average = mean(of Score3-Score10);
run;
```

CALL SORTN switches the values of the variables Score1 to Score10 from lowest to highest. The average of the eight highest scores is therefore the mean of Score3 through Score10.

If you prefer to have the scores in descending order, you can write your call routine line this:

```
call sortn(of Score10-Score1);
```

This way, Score1 will contain the highest value, Score2, the second highest, and so forth. To compute the mean of the eight highest scores, you would code:

```
Average = mean(of Score1-Score8);
```

Thanks to Rick Langston at SAS Institute for this clever idea.

Task: Extracting the first and last name (and possibly a middle name) from a variable containing the first and last name (and possibly a middle name) in a single variable

Keywords

Extract last name

Parsing a string

SCAN function

The last task in this chapter solves a common problem: You have a single variable that contains the first and last name and, possibly, a middle name (or initial).

The way to solve this task is to use the SCAN function to extract each of the names and assign each to a separate variable. This function takes a character string as the first argument and the number of the word you want as the second argument. An optional third argument lets you specify a list of delimiters. Since the first, middle, and last names are separated by spaces in this example, you specify spaces as delimiters.

Here is a listing of data set Full_Name:

Name
Jane Ireland
Ronald P. Cody
Robert Louis Stevenson
Daniel Friedman
Louis H. Horvath
Mary Williams

Notice that some names contain only a first and last name—some names contain a middle name or initial.

You can extract the first name by requesting the first word in the Name variable. The last name will either be the third word (if there is a middle name or initial) and the second word if there is not. If you enter a negative number for the second argument of the SCAN function, the function searches for words from right to left. Therefore, to obtain the last name, use a minus one for the second argument. Here is the program:

Program 12.16: Extracting the first and last name (and possibly a middle name or initial) from a variable containing the first and last name (and possibly a middle name or initial) in a single variable

```
*Extracting the first and last name (possibly a middle initial)
 from a variable containing first (possibly middle) and last name
 in a single variable;
Data Seperate;
   set Full_Name;
   First = scan(Name,1,' ');
   Last = Scan(Name,-1,' ');
   Middle = scan(Name,2,' ');
   if missing(Scan(Name,3)) then Middle = ' ';
run;
```

You obtain the first name by extracting the first word in the string and the last name by entering a minus one as the second argument to the SCAN function. If there is a middle name, it will be the second word. However, if there is no middle name, there is no third word. The SCAN function returns a missing value in that case. So, if you obtain a missing value for the third word, you know there is no middle name and you set the variable Middle to a missing value.

A listing of data set Separate follows:

Name	First	Last	Middle
Jane Ireland	Jane	Ireland	
Ronald P. Cody	Ronald	Cody	P.
Robert Louis Stevenson	Robert	Stevenson	Louis
Daniel Friedman	Daniel	Friedman	
Louis H. Horvath	Louis	Horvath	H.
Mary Williams	Mary	Williams	

Now you have a solution to a very common programming task.

Conclusion

Although this chapter contains an eclectic collection of tasks, they are, nevertheless, tasks that are fairly common. Many of the tasks were presented in several ways—some straightforward, some elegant.

My hope is that this book will help you in several ways: One way is practical—you can copy or modify a task solution to solve a problem of your own. I also hope that by reading this book, I have acquainted you with programming techniques that you may not have seen before.

Index

ACCELERATE YOUR SAS® KNOWLEDGE WITH SAS BOOKS

Learn about our authors and their books, download free chapters, access example code and data, and more at **support.sas.com/authors**.

Browse our full catalog to find additional books that are just right for you at **support.sas.com/bookstore**.

Subscribe to our monthly e-newsletter to get the latest on new books, documentation, and tips—delivered to you—at **support.sas.com/sbr**.

Browse and search free SAS documentation sorted by release and by product at **support.sas.com/documentation**.

Email us: sasbook@sas.com
Call: 800-727-3228

THE POWER TO KNOW®